Toward Solving Forrest Fenn's Hidden Treasure Clues

Plus Research Results on Four Famous Lost Treasures

Revised Edition

Marvin Brooks

© 2016 Marvin Brooks

All rights reserved. No part of this book may be reproduced or transmitted in any form or by any means, electronic or mechanical, including photocopying and recording, or by any information storage and retrieval system, without permission in writing from the author.

First Edition - Revised
ISBN: 978-0-692-71178-1

Designed by Bibliographix

Also by Marvin Brooks:

Researching the Baby Doe Tabor Legend

Researching the Hillary Clinton Scandals

Contents

Introduction	1
Toward Solving Forrest Fenn's Hidden Treasure Clues	11
Researching the Steamboat *Gila* Robbery–Crescent Springs Treasure Story	
Chapter One	37
Chapter Two	44
Chapter Three	49
Chapter Four	54
Chapter Five	60
Researching the Missing Dimes of the Denver Mint Story	71
Researching the Missing Gold From the Steamboat *Far West* Story	82
Researching Maximilian's Lost Treasure Story	98
Conclusion	107

Introduction

Treasure Hunter. That title covers a lot of territory, from those wealthy folks that spend into the millions searching for such elusive treasures as the one they hope to recover on Oak Island, to Mel Fisher, who spent years of his life down in the Florida Keys, with teams of young divers, looking for the spoils from Spanish galleons. He was successful too, in that after sixteen years of searching his crew discovered the Atocha, and reportedly, $400 million in treasure. Yes, Mel was a heck of a treasure hunter, persistent too. I think all that treasure was tied up in a lot of lawsuits; still, I feel sure Mel got his share of it. I've read quite a bit about Mel Fisher, including one book I highly recommend, *Treasure*, by Robert Daley, published in 1977. Unfortunately, there was a tragic event that occurred in those years that Mel spent in his search. Mel had purchased a tugboat so that his crew wouldn't have to make a trip back to land each night; they could sleep on board the tugboat, right over their search site. But the boat was not made for the sea, and one night, even though in calm weather, and without any warning, the tugboat flipped totally over, trapping one of Mel's sons and his daughter-in-law below deck. I believe one or two other divers also drowned. Those few sleeping on deck survived.

But even that tragedy did not stop Mel, and he finally hit it rich, very rich. Now I know many people might ask: but was it really worth it? Were the riches he finally found worth all those years and even the death of his loved ones? I have read enough about Mel Fisher to know some of the answers to that question. Mel stayed at it because treasure hunting was what he did. It was his life. And although I'm sure he enjoyed the wealth he finally found, I don't think wealth was his primary goal. I think Mel Fisher got up each day to enjoy what one of the main characters in this book calls, "The Thrill of the Chase." And I think that's what motivates most treasure hunters, including people that go out with their metal detector in a local park, or pan for gold in some western stream. For certain I was never a treasure hunter the likes of Mel Fisher; rather, I fell in with that latter group. And although I never found anything really valuable, I had some interesting times, and just like a fisherman that thinks that today he (or she) might catch that prize fish, I definitely enjoyed the chase. I have a few examples.

Marvin Brooks

Many years ago, in my working life, I was stationed in the Rio Grande Valley, and spent a lot of time in the Brownsville office of my agency. I learned that one of my friends in that office was from a well-known family of early settlers in that area and in fact still owned property down close to the Gulf that included Palmetto Hill. Palmetto Hill? Frankly, I had never heard of the place, but I soon learned that it was on that hill that the last battle of the Civil War took place, on May 13, 1865, and more than a month after Lee's surrender at Appomattox.

When I first heard the story I assumed that the Confederate soldiers in that battle just hadn't received word of Lee's surrender. Not true. They knew alright, but the Western states of the Confederacy had not officially surrendered, although the governors of those states were negotiating a truce even as the battle of Palmetto Hill (sometimes called Palmito Hill or Ranch) was being fought. The story of this last Civil War battle is an interesting one, and for those of you who want to know more about it, I refer you to an entry on the internet by the Texas State Historical Society, entitled, "The Battle of Palmito Ranch." It's a good one. And I can tell you the Rebels won that battle hands down, incurring few casualties while wounding or killing thirty of the Union forces. It was a battle that should never have been fought, and from what I gather it was caused by the arrogance or ambition of one or two Union officers, who initiated the battle. Again, I refer you to the internet.

But I need to get back to my "treasure hunt" out there on that hill. My Brownsville friend very graciously loaned me keys to the gate of his property, gave me directions, and I was on my own. What did I expect to find? Well, together on both sides, 1,000 men fought in that battle. Isn't it possible that they left some artifacts on the battle field? Maybe dropped some coins, even gold ones? A pistol? Heck, even a button from a uniform?

Well, that next weekend I spent most of a day out there. As my friend informed me, he had leased out his property, and most of "the hill" was now planted in cotton. I found that "the hill" backed right up to the Rio Grande River, and I gave it my best shot, out there all alone. Up one cotton row and down the other. I found bullets, but I doubt if they

Introduction

were from that battle. There was also evidence of an old homesite, but there was so much metal trash around that it was impossible to search by metal detector. I had a great time, but no success, so I packed it in, thinking I would come back again at a later date, as I thought I might rake up and discard the trash around a good size area of the homesite and see if I could find something of value in the ground without that interference, and also maybe find the trash dump of those long-ago residents. So why didn't I go back? Well, I'm not sure, but I think I had a girlfriend, and, you know, girlfriends trump treasure hunting, almost every time. Many years later, after retirement, I was in the Brownsville area and drove out to Palmetto Hill. The area where I had searched was no longer a cotton field—there was a mobile home on the site, but still some vacant land. I thought maybe my old friend lived there, but the home was fenced in and locked. I couldn't go up to knock on the door, but stood outside and hollered his name loudly. But no one responded so I went on my way. If any of you readers are ever out that way, there is a historical monument giving a brief history of "the battle," and a picture of that monument is avalable on the internet.

    I also had another interesting experience while living in Brownsville, having to do with the history of that area and "treasure hunting." In truth, I was not searching for "treasure," I was looking for artifacts, but what the heck, to me, anything found, of any value at all, constitutes "treasure." My definition of "treasure" is very generous. But to my story.

    I learned that there had been two communities at the mouth of the Rio Grande back in the Civil War era: Clarkville, in the U.S., and Bagdad, right across the river in Mexico. Now Clarkville was no problem for the Union, they occupied that little town throughout the Civil War, but Bagdad was a different proposition. Mexico was a neutral country, and Bagdad offered the Confederacy a major port to export cotton, and to import needed goods. Any goods, imports or exports, were brought to and from Bagdad to Matamoras, and crossed the Rio Grande at Brownsville, held by the Confederate forces at Ft. Brown. Of course, all of this relates to that battle discussed above. The Union

Marvin Brooks

wanted to put a stop to that situation by attacking and defeating those Ft. Brown forces.

Now Bagdad continued as a major seaport even after the Civil War ended, that is, until November 7, 1867, or about two years after the war ended; then a devastating hurricane hit at the mouth of the Rio Grande, which virtually destroyed both towns, and also the community on Brazos Island. After the storm very little remained, and many perished, but I don't know the exact number. For a good account of this tragedy, including first-hand accounts by survivors, I recommend a website, *Gendisaster.com*, "Bagdad and Clarkville." But in the end Clarkville rebuilt, and continued on for a few years until further storms hit, then it passed into history. Bagdad never recovered from the storm, and, from the reports I have seen, nothing at all remains of this once-important seaport. And that might not be a bad thing. Here is one description of its inhabitants: criminals, cutthroats, pirates, smugglers, renegades, and of course, prostitutes, lots of prostitutes. It was considered the wickedest city on earth during its day, and its residents were considered "Scums of the Seven Seas." It must have been quite a place.

So, hearing about all this, I headed out one day to the mouth of the Rio Grande. It would be easy to go from the main road to the destination with a four-wheel drive vehicle, but I didn't have one, so I walked, probably about a mile. I saw that the mighty Rio Grande, that starts in the Rocky Mountains of Colorado, goes throughout the state of New Mexico, and provides a border between Texas and Mexico for the entire length of that border is not so grand where it finally empties into the Gulf. It doesn't roar out to sea, it dribbles out, and Mexican children were wading in it—Mexico to Texas, Texas to Mexico; well, you get the picture. That was more than twenty-five years ago—today, I wouldn't even venture down to that area. If I came back with my life, or even my metal detector, I would probably be lucky. But back in those days, things were different. I had no fear at all—saw nothing the least bit threatening, and I would cross the bridge into Matamoras with absolutely no concerns about safety. Not so nowadays—so if any of you readers decide to go to the mouth of the Rio Grande to search for valuables blown away in that

Introduction

1867 hurricane—beware—be careful—and don't even think about going there as I did, alone and unarmed. But what about the "treasure hunt?" I know the question: "Did you find anything?"

I learned many years ago that when I tell someone about going out "treasure hunting," or prospecting, the first question they always ask is, "did you find anything?" Meaning, did you find anything of value. Now I've heard other fellow "treasure hunters" really lay it on thick, saying something like this: "Yeah, I spent two weeks out there on the California coast. Man, I was up at daybreak out on the beach, and in the surf, with my metal detector. I mean I was out there all day, for two solid weeks, combing those beaches." "Did you find anything?" "Did I find anything? Five rings, four necklaces, three bracelets, and over a hundred coins!" One can see the mind of that questioner starting to come alive, as he (or she) is thinking, damn, I've got to get one of those metal detectors; and some ask, how much does one of those metal detectors cost? But what that braggart forgets to add is that most of that jewelry was lost by little girls, and most, if not all, that jewelry could be bought at the local Goodwill store for two dollars. Well, that's not me. When someone asks me if I found anything I just smile mysteriously and say, "well, I did all right, but I don't really like to talk about it, the IRS, you know." Of course, my friends know I am probably lying, but it does create some doubt, and gets them thinking, "say, how much do those metal detectors cost?" And so I'll leave it at that as to what I found down there where the Rio Grande meets the Gulf of Mexico—the Texas Antiquity Act, you know. But can you imagine what it would be like in that same location down around 1865—I'll bet I would have heard those low-lifes and scumbags in Bagdad all the way over to Clarkville, just raising hell. Well, enjoy it, boys, because November 7, 1867 is just around the corner.

Somewhat luckily I have been retired for a couple of decades now. I will admit that I was much more interested in low-limit poker during most of those years rather than going out on the beach or into the desert with my metal detector, or up into the Rockies with my gold pan, along some mountain stream. Note that I said low-limit poker. That's the kind where you can sit down for hours and play without fear of any life-altering losses, and there is some conversation and bullshit going on, and I've sure enjoyed the competitive nature of poker, even the low-limit kind. But every now and again I would get burned out—then I would think of grabbing the detector and taking off.

So it was that one autumn—I had been playing poker that summer up in the mountains at an old mining town, but now a little gambling resort named Cripple Creek—I decided to head out to the Eastern coast of Florida, the famous Treasure Coast, where I always wanted to beachcomb—looking for those gold doubloons so many treasure hunters have found. Fall is the right time for a person like me to be in that area. The summer crowds are gone, and the winter tourists won't be arriving until around Christmas time when prices down there just about double. That was before I bought an RV, so I stayed in those weekly rentals, as I traveled to a number of locations starting in Daytona Beach and ending almost to Palm Beach.

Florida, like most states including Texas and California, has an open beach law. That means that a "Joe Citizen" such as myself can enjoy the same beach as the billionaires of Palm Beach. Okay, there is one problem. It's not always easy to find an "access" to those type beaches, and once you find one, you might have to walk a spell to be on that "public" beach shared by the rich folks. I might add that I made a number of these walks, and I saw very few people I thought were from up above in the beautiful homes. I finally concluded that they were too busy up there making money—probably on the phone with their broker—and I felt really privileged not to be so encumbered. Yeah, right.

I walked those Florida beaches in the early morning, went home for a lunch, and returned in the late afternoon. I can truthfully say that no one paid the least bit of attention to me. I searched the beaches and I searched the dunes. Occasionally, at low tide I waded out a bit into the surf, but one must be careful unless the detector is waterproof and made for salt water. Those Atlantic waves come in strong from time to time and I barely escaped being swamped a few times. So you ask, did I find anything? Yes, coins, lots of coins. I also found some interesting

Introduction

bits of metal—strange looking. I still have them. Bits off the crash of the Challenger? Not likely, but who knows?

I once met a fellow detectorist out in a park, an old guy. He told me that he had just found his 10,000th coin, and had kept every one of them. But one rainy day in Florida, with nothing to do, I took out all the coins I had found and counted them, then figured their entire value, giving me an average value per coin. Okay, I'll admit there were a bunch of pennies in the mix, but I came up with an average value per coin of seven cents. Now I'm sure the old guy found a lot of coins that were not clad, and therefore worth more than face value, and although I didn't ask him, I'm sure he found some jewelry along the way, maybe something more than little girl rings; but if those 10,000 coins were all clad, and using my "empirical research" findings, they would be worth $700. I forget if he told me how long he had been at it, searching for coins, but I got the impression he had been at it for many years. And no telling how many hours he had put into his particular endeavor. Why, I'll bet he could have earned much more money at some fast food place. But folks, I'm just being silly. What he was doing is a great hobby, and it's obvious that he got the same enjoyment from his hobby as anyone, including fishermen, and we can be sure that each day he went out he expected to find that big gold necklace, or multi-carat diamond ring. How do I know that? Because I was, at one time, of that same mindset, and still would be if my health allowed it. No, I didn't find any old gold coins washed up on the beach from some sunken galleon. But I had a great time on the Florida beaches that fall, and only left when the snowbirds arrived and I could no longer afford Florida's winter season costs. "What do you expect?" they say, "It's the season." But I came out well on top, I had more fun than those rich folks up in the mansions. Well, maybe I did.

During my years of retirement, I finally bought an RV, and lived exclusively in that class C, twenty-six footer for seven years, until an inconvenient heart attack forced me to settle into an apartment. But during those seven years I spent the summers parked somewhere around Colorado Springs, where I have family. From time to time, when I got bored, I would take off into the Rocky Mountains, intent on panning out some of those gold nuggets so plentiful in those streams up there. Well, aren't they? I had a few laughs up there. Once I was up at Fairplay, panning for gold on the upper Platte River. My method was to fill up a couple of five-gallon paint buckets with "very promising" sand and gravel, then sit by the stream with my wading boots in the water, and pan out that gold. I wasn't having any luck. Then here came two carloads of folks, and when they got out I noted a couple of the menfolks were studying a guide book which apparently told them they were at a prime location for finding gold. Now with my wading boots on, surrounded by my "tools," and with my grizzly-looking beard, I probably looked like a seasoned old prospector, because one of the women in the group said, "Oh, I'm so glad there is someone here that knows what they are doing." I almost fell into the river laughing—I think that was my second trip out "prospecting." They turned out to be a great bunch of folks.

Another time I was up on the Arkansas, camping in a state park. I was doing my same routine, filling up paint buckets and panning out on the river bank. I just wasn't having any luck, which really got to me because a fellow "prospector," just a bit upstream from me, showed me in a vial filled with water containing the gold he had panned out over a two-day period. I would estimate that there was two or three dollars worth of gold there, but still, it was gold, and I wasn't finding any. I decided I needed to fill my buckets mid-stream, and the Arkansas wasn't that deep where I was. So with my bucket and GI shovel, I waded out. I would probably have done alright, except I tried to walk too fast against the current, which was stronger that I thought it was, and I stumbled. The next thing I knew, I was completely underwater, and I remember very clearly what my thought was: "What am I doing here?" My hat and paint buckets floated on downstream but I was able to retrieve everything. My wading boots were filled with water. But luckily I had my RV there—changed clothes—and reassessed my "prospecting" methods. I decided that it would not be a good idea to wade out into the current for the rest of the trip.

I once thought it would be a great idea to travel around the complete shoreline of the U.S., beachcombing and metal detecting

Introduction

on the beach and in the surf. I would be sure and have a waterproof, saltwater-safe top notch detector. I would start up on the northernmost seashore of Washington State, follow the coast through Washington, on down through Oregon, and Northern and Southern California. Then I would head to Brownsville, Texas and follow the Gulf all the way from there to the Florida Keys. Then I would go all the way to the uppermost seashore of Maine. I figured that the trip woud take about two years. Now mind you, I planned (in my mind) that I would search the beaches for treasure along the entire coastline of this country. Of course I would write of my experiences as I went along, then publish a book entitled, "The Adventures of the World's Greatest Beachcomber." What a great best seller that would be. But the truth is, I knew I could never make that trip—not alone. It's just too lonely out there, and I know I would never find a woman to accompany me. In a Class C RV! I remember the salesman that sold it to me said, "you know, if you were married you would be right back in here, trading this in on one of those big Class A's." Why? "Because there is just not enough room in that rig of yours for a woman to do the things she wants to do—why, there isn't even enough closet space for a woman." And I think he was right. Some special types of women travel in a Class C, but I don't think many of them live in one permanently, as I did. Besides, only a trip of one month proved to me that I got too lonely camping out on a beach. It was different in Florida. I was living in apartments and eating in restaurants. There was always someone to talk to, if only briefly.

But regarding the loneliness, I very well remember a trip I took along the Southern California coastline. I was bored, I thought, with the poker routine in Laughlin, Nevada, and the plan was to start my camping at the first California State Park north of San Diego, stay in each park three or four days, then head on up the coast to the next one. In those days, seniors were given great discounts, and I had it all planned out. It worked perfectly, for one month. In that one month I stayed at nine state parks, all very pleasant. I had no problem getting through Los Angeles, and kept going, making reservations at the next park on the agenda, always a few days prior to my arrival. I believe my last park was Big

Marvin Brooks

Sur. Then I planned on going around the Bay Area, and continue up the Northern California coast to Oregon. After that, I wasn't sure if I wanted to go further. But by the time I reached the San Jose area, a great sense of loneliness overcame me. I recognized that, while I was enjoying the trip, it seemed rather meaningless without someone to share the experience with me. I turned back south, not the way I came, but Highway 101, heading back to Laughlin, Nevada, and my poker game. I've thought about that trip. Was I, in fact, addicted to poker, to gambling, or did I abort my trip due to loneliness? Probably a little of both, I think.

    I think that is enough of my beachcombing and "treasure hunting" experiences, although, as you will see, one of the five stories I present in this book is my research experiences in tracking down the origins of the Steamboat Gila robbery and the lost treasures of Crescent Springs, Nevada, a treasure story that I really got caught up in. I thought about featuring that story in this book with a title to match. But that story is not well known. Why not try to solve at least part of the Forrest Fenn clues, and feature that story in my book? I decided on that approach, because, let's face it, no other treasure story, with the possible exception of that Superstition Mountain legend, has gained as much public attention as Fenn's hidden treasure. I believe Fenn himself said that 30,000 people had actively searched for the treasure, and I feel sure many thousands more have tried to solve his great puzzle. I will admit, the poem, the clues, are much more difficult than I thought they would be. Good maps, a little logic and good common sense would, I thought, at least aim a seeker in the right direction. I've tried to use those three tools, but I am far from certain that I solved any of Fenn's clues. However, I trust my efforts will be worthwhile reading for the Fenn treasure hunter, and I hope that my research efforts into the other famous, but probably myths, will also be of interest to anyone holding the self-named title: Treasure Hunter.

Toward Solving Forrest Fenn's Hidden Treasure Clues

I suspect that anyone claiming to be a treasure hunter has already heard plenty about this strange treasure story. But here it is in a nutshell: Forrest Fenn is an Air Force retiree who, after retiring, moved his family to Santa Fe, New Mexico, and subsequently opened an art museum, which proved to be very successful. But around 1988 Fenn developed cancer, resulting in the removal of a kidney. He didn't think he would survive, but he did. Perhaps it was this traumatic event in his life that led him to start thinking about hiding a treasure, and providing clues to the finding of that treasure in the form of a poem in a book, his memoir. I've seen where he contemplated all this for many years. I think most people at this point, not real familiar with Fenn's story, might ask, why would he do that? Well, I think Fenn reveals in his book why he hid that treasure, "for all to seek," and I'll discuss that later in the essay.

So in his book, *The Thrill of the Chase*, he tells of hiding that valuable antique bronze chest filled with gold nuggets, rare coins, jewelry, and gemstones—things of value that he had collected over the years, and he also included a 20,000-word autobiography, to tell the finder something about the person that hid the treasure. He refuses to give an estimate of the value of his treasure—pointing out how it will fluctuate over time. But it has been estimated by some "experts" as worth between one and three million dollars, and maybe more. Who knows for sure? And in his book, published in 2010, Fenn offers his poem, twenty-four lines, which he claims provides nine clues that will lead the reader to the treasure. Well, heck, that sounds easy enough, where can I get a copy of that book?

Not so fast. Remember, those unspecified nine clues are hidden in a twenty-four line poem. Perhaps Taylor Clark, in his article for *The California Sunday Magazine* entitled, "The Everlasting Forrest Fenn," describes the poem best. He says it is "Twenty-four lines of mind-boggling ambiguity, obscurity, and general vagueness." This article by Clark, with great pictures of Fenn, including ones of Fenn's library and his collections, by Jesse Chehak, is available online. I also want to add that there is another great article available online by Mary Caperton

Marvin Brooks

Morton entitled, "On the Trail of Treasure in the Rocky Mountains," from *Earth* magazine, dated 2-15-15. These two articles are must reading for anyone interested in the Forrest Fenn hidden treasure story. But with Clark's description fresh in our minds, here is the poem:

> As I have gone alone in there
> And with my treasures bold,
> I can keep my secret where,
> And hint of riches new and old.
>
> Begin it where warm waters halt
> And take it in the canyon down,
> Not far, but too far to walk.
> Put in below the home of Brown.
>
> From there it's no place for the meek,
> The end is ever drawing nigh;
> There'll be no paddle up your creek,
> Just heavy loads and water high.
>
> If you've been wise and found the blaze,
> Look quickly down, your quest to cease.
> But tarry scant with marvel gaze,
> Just take the chest and go in peace.
>
> So why is it that I must go
> And leave my trove for all to seek?
> The answers I already know,
> I've done it tired, and now I'm weak.
>
> So hear me all and listen good,
> Your effort will be worth the cold.
> If you are brave and in the wood
> I give you title to the gold.

Toward Solving Forrest Fenn's Hidden Treasure Clues

That's it—simply decipher the poem and go get your treasure. But wait! Fenn has given us a few more clues in the years since he first published his poem. He tells us the treasure is at least eight miles north of Santa Fe, and in the Rocky Mountains, at an altitude of over 5,000 feet. Then later, after people were getting in trouble climbing high mountains thinking they would find the treasure on a mountain top, Fenn tells us that it is at an elevation under 10,200 feet (and that's pretty high). Then, after the treasure seekers started to destroy buildings in their "frenzied" chase, he lets us know the treasure is not associated with any structure; and stay out of the graveyards, it not in there either, he tells us. Then Fenn, in his majestic mercy, reveals that the treasure is not in either Utah or Idaho. Well, now we're narrowing this thing down. We only need to look in the Rockies in Montana, Wyoming, Colorado, and north-central New Mexico, where the Rockies play out. Fenn also tells us the chest is not in a cave, and that it is wet. So let's take a close look at that poem and see if we can solve those clues. Then, with all those additional clues Fenn has so generously given us, this should be like a walk in the park.

I should point out that this is not the first time I've delved into the mysteries of Fenn's treasure and made an attempt to decipher his poem. A couple of years ago, I made a half-serious effort to find that hidden treasure; borrowed a copy of his book, and tried to make sense of his poem. But I won't say I gave it my best shot. Oh, I spent quite a few hours with the poem, and maps of the Rocky Mountains, New Mexico, Colorado, and particularly the Yellowstone National Park, where Fenn tells us he spent many happy summers during his youth, on long vacations with his family. Like other "seekers" I noted that there was a "home of Brown" in the park, the previous home of retired park ranger Gary Brown. The location is now known as Lamar Ranger Station. I thought that location might be the "home of Brown" in Fenn's poem, the only reference in the whole poem to a place-name, or a capitalized word, except for the first words to a sentence, and I think most people studying the poem for clues believe those words—home of Brown—is the key clue in the poem. I should point out that "the home of Ranger Brown" would not be part of Fenn's childhood memories. Ranger Brown didn't go to work in Yellowstone until 1965, at which time Fenn was 35 years old.

But I couldn't connect those words, home of Brown, with any other possible clues that made any sense to me, although there are many physical features in Yellowstone: canyons, creeks, warm water, high water, etc., that might relate to the poem, so I finally gave up on the Yellowstone. One big reason is this: the management of that park is well aware that people are coming into the park looking for Fenn's treasure. Apparently, some have been destructive, and the park people have made it clear that taking any property out of a national park is a serious crime, even if it was hidden there by some 80 year old, sometime about year 2010. So I trust that Fenn, being familiar with the Yellowstone Park, would not hide it there and risk sending the finder of the treasure to jail. Besides, those greedy Feds would no doubt confiscate the treasure chest and claim it as their own property.

Now, looking back to my previous effort in trying to figure out Fenn's clues, I thought about Colorado's own Arkansas Headwater Recreational Area, which was much closer to my home in Colorado Springs than Yellowstone Park—only about seventy miles west. And that area is right in the middle of the Rockies. Furthermore, there is a very well-known canyon, especially known to rafters, the Brown Canyon, and a little ways on down, a Brown Creek, and Brown's Creek Trail, and there might be a place right in that area "where warm waters halt" because there are hot springs, a tourist attraction, not too far away by Mount Princeton, and a creek takes water from those springs down to the Arkansas River, which usually runs cold. Well, I just spent a lot of time trying to connect some of the "clues" in Fenn's poem to the AHRA. I knew from my research back then that Fenn was known as a crafty old guy that might very well use double meanings to confuse us, the seekers. And I knew that many people think the word Brown, capitalized, is a name that very possibly refers to brown trout. Further, Fenn is known to be a fisherman, possibly a serious fly-fisherman. The Arkansas River, right there at Brown Canyon, is known for its good brown trout fishing. Of course, some "searchers" believe that the term, "home of Brown,"

Toward Solving Forrest Fenn's Hidden Treasure Clues

is a Fenn clue to any place in the Rockies where there are brown trout, and is especially pertinent to any river, stream, or creek that fisherman Fenn might have considered his "favorite fishing spot," if one could determine where that might be. Now looking at the map, I could see where a major highway, 285, goes directly from Santa Fe north, all the way to the Arkansas Headwater Recreational Area. Then I saw where Brown's Creek empties into the Arkansas River, and nearby is Brown's Creek Trail, which, as I learned, leads to an impressive waterfall. Could Fenn have hidden his treasure along that creek, or even behind or somewhere around that waterfall, and could there be a "blaze," which might be a marked trail leading a "seeker" directly to that treasure?

I did indeed think of going over there to check out those possibilities and clues in the AHRA. But it was winter; I would have had to wait until the spring, and I had other plans. And there is something more. I really didn't think my "solutions" made a lot of sense, you know, like speculating that Brown's Canyon was the "home of Brown," or that the point where the creek from up around the hot springs of Mt. Princeton emptied into the Arkansas was "where warm waters halt." And one other thing, I could find no evidence that Forrest Fenn ever had any ties to that Colorado park, or ever mentioned it in his book, as he did the Yellowstone or areas closer to his home in Santa Fe. What I really learned from all this is how easy it is to fit some of Fenn's clues into whatever location one chooses. But even at that, these "solutions" that I have related here make as much sense as some "theories" on Fenn's treasure that I have read about, and some of those "theories" have even resulted in books.

So why am I coming back to this most frustrating of treasure stories? Well, I've since learned that someone such as myself, and probably anyone buying this book, with the mindset of a "treasure hunter," might well be bedeviled by Fenn's poem forever, and I begin to think that the answers to some of his clues might be akin to crossword puzzles in that the answers, many times, are simpler than the clues. Surely, if I put my mind to it I could interpret those clues. Also, I will admit, I wanted to publish a book on researching strange treasure stories such as the one featuring my extensive research on the Steamboat Gila–Crescent Springs treasure tale, and I think anyone would consider Fenn and his hidden treasure a strange story worth pursuing.

My first thought was that I would need to solve each one of these clues sequentially, so that if I could solve clue #1, that would lead to the next clue, which would then need to be solved, and so on, until the final clue, solved, would lead to the treasure. I think that most Fenn treasure seekers follow that line of thinking. I tried it, and soon learned that that is probably an impossibility. I finally determined that I would never solve Fenn's clues by using a "sequential" approach.

So what is the answer? I concluded it would be a series of what I call "leaps of faith," based on the best logic and information I could bring to the task. I recognized that these "leaps" would not always be correct—maybe never—but I thought this approach must be better than the others, where people can not even determine which of the Rocky Mountain states the treasure is located in. So what follows is an account of my research efforts, where I think a little reasoning, imagination, information, and a few of those "leaps of faith" might lead the Fenn treasure seeker to a general location of the treasure, while fully recognizing that it will take an on-site "exploration" to locate those "blazes" or whatever signs Fenn has provided to finally lead a "seeker" to his treasure. In other words, I don't think his clues will lead one directly to his treasure. So we must first find its GENERAL location. Of course, I might be wrong, might be totally off track, but I believe my interpretations and speculations point to that specific general location. Let's look at the clues.

I recently re-read *The Thrill of the Chase*. This book is probably not what most people would expect in that it's a big book, 12 inches by 8.5 inches and 147 pages; obviously designed to remind readers that it is something of a family album. It has many, many photos of people and things, such as airplanes, that are important to Fenn. It's a memoir of Fenn's life, and his treasure story doesn't enter the book until one of the last chapters. But this time I read the entire book, and I think I learned a few things relevant to Fenn's treasure, and Fenn himself, even if I failed

Toward Solving Forrest Fenn's Hidden Treasure Clues

to find some all-important clue to lead me directly to that little chest. In writing about his experiences growing up, he displays a kind of self-effacing sense of humor. His stories about his "battles" with Miss Ford, his middle school teacher, in a school where his dad was principal, or his escapades with his brother Skippy, are hilarious, and I found myself laughing out loud at some of them, rare for me. I also enjoyed his chapters on his flying experiences in Vietnam. Fenn flew 274 missions in that war, and had to bail out and be rescued twice. This man is a true hero. I was also impressed that he joined the Air Force as an enlisted man, applied for and was accepted into pilot training, emerging as an officer and flyer of jet aircraft in Vietnam. That's quite impressive, as is his success in the art business. But what about the parts on his treasure? Anything there of value to us searchers? I think there is, and I think what I learned helps us understand two things about Fenn that might be important to us in understanding at least two of his poem's stanzas, #1 & #5, in effect telling us what he did, and why he did it. Let's start with stanza #1.

    One of the most important things I got out of Fenn's book, right behind his love of family, is his overriding admiration of courage. He alludes to this a number of times in his book, both directly and indirectly. So in looking at stanza #1, he makes reference to courage in his very first sentence when he says he went in there (where he hid his treasure) alone, and bold. Bold is a synonym for courage, and I think Fenn is telling us that it took courage to do what he did, which implies that his treasure might very well be hidden in a dangerous place. Also, in the very last sentence of his poem he tells us we will find his treasure ONLY if we are Brave. I think we will find that it will probably take some boldness, bravery, or courage, they all mean the same, to go down in the area where I think Fenn's treasure is hidden. So to understand stanza #1 of Fenn's poem, I think we need to understand his "admiration of courage." Further, I think he is telling us in stanza #1 that he will keep the secret (not share the location with anyone) and will provide hints (clues) as to where the chest is hidden—that chest being "riches new and old." I might also add that Fenn admires an "adventurous spirit" and invokes a familiar name that has that spirit—Indiana Jones. Now let's look at stanza #5.

Marvin Brooks

I could never understand anything in that stanza until I carefully read Fenn's book. But Fenn spends several pages discussing one particular theme that is perhaps the most eloquent and philosophical part of his book. That is, and I'm summarizing, most men (and I'm sure Fenn also includes women), no matter what they achieve in life, no matter their heroism, or their contributions to the world, are soon forgotten. After making a very eloquent case that all but the greatest of men are soon forgotten, Fenn says, "But what about those of us who are not great men? Are we not entitled to leave a slight footprint... somewhere?" I could never figure out what Fenn is saying in stanza #5 until I carefully read this in his book. Then in his poem he wrote, "So why is it that I must go and leave my trove for all to seek? The answers I already know," he then adds, "I've done it tired and now I'm weak." Now I might be totally wrong, but I think in that stanza Fenn is telling us that he recognizes that he is growing old, that his life is coming to an end, and that his hidden treasure is, in effect, his "slight footprint," a way not to be forgotten. Fenn has said his treasure is "difficult to find, but not impossible." That being the case, the world might well be searching for that little chest for the next hundred years or more, and it might become a part of our folklore. And who can blame Fenn for not wanting to be forgotten? Now let's turn to the difficult part, trying to unravel Fenn's clues and at least find a reasonable "solution" to finding that general location of his treasure.

When I first started trying to find a likely location for Fenn's treasure, I bought a big map, 3.5 feet by 4.5 feet, from the U.S. Forest Service, for northern New Mexico, and combed that map for any location I thought might fit Fenn's clues. I found the map to be overwhelming—there are just hundreds of canyons and creeks that could be a hiding place for that tiny chest. Then I tried to focus on the line, "Begin it where warm waters halt." I figured IT meant "the search." So where was WWWH? Some searchers, I learned, believe that WWWH was a specific place in the Rockies where warm water ran into cold water, but I couldn't find anything on that New Mexico map where I thought that term applied. Then I learned somewhere that WWWH is a term used by trout fishermen. I noticed one thing. If I was right in my belief that Fenn's treasure was hidden in northern New Mexico, in the area considered to be in the Rockies, and close to a river known for its trout fishing, and WWWH, one river stood out, the Chama.

After a lot of thought and a little reseach I concluded that "Where warm waters halt," does not refer to any one particular spot anywhere in the Rockies, but refers to any trout stream, specifically in northern New Mexico where waters, too warm for trout, become cold enough for them to survive. Northern New Mexico, the home of Fenn, is well known among fishermen as a place WWWH, and as I said, the Chama River stood out. So I started on some research of that river. Here are a couple of items from that research.

A wonderful website for a fishing guide business in Santa Fe named, *Land of Enchantment Guide*, offers this description of the Chama River:

> The Chama River is probably one of the West's most undiscovered and diverse trout streams. Starting in the mountains on the Colorado border, it flows as a freestone stream for many miles through high mountains, forests and meadows. Below El Vado reservoir, it runs for over thirty miles through rugged, multi-colored, sandstone canyons and rough terrain with limited access. For six miles of this section it passes through the 50,000 acre Chama River Canyon Wilderness Area. Another less remote stretch flows below Abiquiu Dam, down to the Rio Grande. All the parts of the Chama have good populations of large wild brown and rainbows.

Of course, I didn't know what a Freestone stream was, but an entry on Wikipedia explained it: "In fly-fishing a FREESTONE stream flows seasonally, based on the water supply. In the summer and fall, the freestone streams grow WARM and have reduced flow because water from snow melt is less readily available. Limestone streams are usually fed by springs, providing cooler water, while freestone streams are supplied by snow melt and runoff."

So based on a "leap of faith," I made the conclusion that the general search area was in northern New Mexico, and specifically on or near the Chama River. Based on my further research there were a couple of other reasons for my conclusion. We know that Forrest Fenn is a fisherman, and probably a serious fly-fisherman. The Rocky Mountains come into New Mexico on a slant from west to east. Therefore, looking northwest from Santa Fe, toward the Rockies, the area is dominated by the Chama River. Also a main highway, Hwy 84, goes north from Santa Fe on up to the Chama River area, providing Fenn with an easy access to that river. Any other reason for zeroing in on the Chama? Well, I wondered where the New Mexico record brown trout was caught. That would be the Chama River. Weight: 20.5 pounds. Length: 35.5 inches. The point being, why would Fenn fish in any other river, and doesn't it make sense that he would have found the place to hide his great treasure while on a fishing trip? Makes sense to me. There is at least one other reason why I believe the treasure is hidden in northern New Mexico. In the great article by Mary Caperton Morton referenced above, she provides this quote: "Based on the numbers of emails he's received, Fenn estimates about 30,000 people are looking for the chest, but, as far as he knows, nobody has found it yet." Now whether or not Fenn's estimate is correct, there are a lot of people out there looking. And I believe most of those "seekers" are looking in northern New Mexico. As a matter of fact, I have seen where this is becoming something of a tourist attraction. Could all those people be wrong? They surely could be, but if anyone finds the treasure in some other location, there will be thousands of really "peeved" people, and Fenn will probably have to leave Santa Fe. So I'll stick with my reasoning—northern New Mexico and its Chama River area, home territory of Forrest Fenn, offer the best bet as a very general location of that hidden chest.

Looking at the clue "take it to the river down, not far but too far to walk," I believed that made sense in that I would find, along the thirty-mile stretch of the Chama from El Vado Dam to the Abiquiu

Reservoir, a stream or canyon entering into the Chama that would fit the first three lines of stanza #2, and I concluded that the fourth line was just a term referring to the Chama River, that river being the "home of Brown trout," as I could not find any other meaning that made sense, in all of northern New Mexico. But where is that stream, that creek, with "just heavy loads and waters high," that fits stanza #3? Using my "Big Map" I started focusing on that area.

At this point it was finally brought home just how difficult, if not impossible, it would be for this mere mortal to ever decipher enough of Fenn clues to ever lead me to his treasure in this area. The map revealed that the "search area" has literally countless rivers, canyons, and creeks, any of which might be that creek where Fenn hid his treasure. Seriously, I studied that big map diligently to see if I could fit Fenn's clues into that area from El Vado Dam to Abiquiu Reservoir, along with a free map the U.S. Forest Service sent me that was very helpful, entitled, "Rio Chama: Wild and Scenic River." The maps indicated that there is an imaginary boundary all along the Chama River called the "Corridor." Along this boundary there are seven canyons: the Aragon, Dark, Mine, Huckbay, Chavez, San Joaquin, and Ojitos. I listed those names to show that they have no obvious relationship to Fenn's clues. Also, I need to point out that I translated all of the major words in Fenn's clues into Spanish and found nothing of interest, not only on the Chama but nowhere on the big map, with one exception, which I will get into further downstream. Brown, by the way, translates as marron—nope, nothing there either. Very discouraged, I gave up on that thirty-mile stretch in looking for the general location of Fenn's treasure.

At this point, I clearly needed to make another "leap of faith," and I did, starting way down in the last sentence of Fenn's poem where it says, "If you are brave and in the wood I give you title to the gold." Those words, "in the wood" always struck me as being odd. If Fenn meant "in the forest." why not "in the woods." Now I thought from the beginning that some of Fenn's words might have some meaning on the maps if they were translated into Spanish, not too far-fetched for northern New Mexico. I translated a lot of those Spanish names on the map to English and some of Fenn's words to Spanish, and found nothing that made any sense, until I translated "the wood" to Spanish, and came up with "La Madera." So is there anything of interest on the maps named La Madera?

As it turns out, there is. Please refer to the maps I have provided. There is not one trail, but two, designated 103 & 104, named La Madera Trail. These trails are very near the "Poshuouinge Ruins," a tourist attraction in northern New Mexico. These trails lead down to La Madera Arroyo. Could La Madera Arroyo be the general location of Fenn's treasure? Do the clues fit? Look at the line, "There'll be no paddle up your creek," and that entire third stanza. It seems to me that Fenn is refering to his hiding place as being in or close by a creek with "heavy loads" which is a term used by watercraft folks in reference to a stream or river not suitable for their sports because of big rocks or other obstacles in the water. Now the word for creek in Spanish is arroyo, and it's also commonly used in the U.S. as a name for a gully, or canyon, usually caused by swiftly flowing water, high water, especially during flooding. So I'm speculating that Fenn's term "in the wood" refers to the Spanish translation "in the La Madera" and that La Madera Arroyo perfectly fits Fenn's third stanza. Further, I am surmising that this location fits stanza #2. We begin our search at the Chama River, if that refers to WWWH, and take it to the "canyon down" meaning southward or toward the bottom of the map. Also, looking at those trails on the map, I can clearly see where the line in stanza #3, "The end is ever drawing nigh" would fit that trail. I might also add, that in looking at the satellite maps, the La Madera Trails appear to be "no place for the meek."

But what about "Not far, but too far too walk," does that fit? Well, the "big map" has a legend indicating 1.25 inches equals one mile. I have studied Google Maps of the La Madera Trails. They twist and turn. I estimate it is at least six miles, maybe more, from the trailhead off Hwy 84 down to La Madera Arroyo, meaning at least twelve miles round trip, and that's "too far to walk." I will admit that I'm pretty well stumped on the interpretation of "Put in below the home of Brown." "Put in" is another term used by the watercraft people, meaning the

Toward Solving Forrest Fenn's Hidden Treasure Clues

place where they "put in" their watercraft. I can only guess that "the home of Brown" also refers to the Chama River, home of brown trout, below which we are to "put in" our vehicle—on one of the trailheads off Hwy 84. But there is still another question—is it possible to drive down to La Madera Arroyo?

Now I do not know if this is a clue in Fenn's book, but he does make a statement about his own vehicle preference, a pickup. If it is a clue, he is telling us that we don't need anything more than a pickup to go to the places he goes. Well, maybe a good four-wheel-drive vehicle would be handy. I should point out that the "big map" clearly indicates that trails 103 and 104 are for "non-motorized use." But in studying those trails on Google satellite maps, I learned that the reality of those trails bears little resemblance to the big hard copy map I was so proud of. I also learned that I could not get a real closeup of the trails or of La Madera Arroyo, from Google maps—finally the maps just "blur out." So to conclude, I would say it's not certain that one can get down to La Madera Arroyo by regular vehicle, but I would say it's probable.

Finally, I should point out what I found surprising. One can simply key in La Madera Arroyo on the internet and there you will find a lot of information, such as a number of maps, and the fact that this arroyo is referred to as a stream, obviously meaning there is water there. Also, that it is at an altitude of 6,400 feet, in keeping with Fenn's statement that his treasure is located at an altitude over 5,000 feet.

Alright, so here we are—brave souls—no, we didn't come to the Arroyo alone, we came with a group "expedition." But we're here, now how do we find that treasure? Good question. Are there any clues?

It seems to me that starting in stanza #4, Fenn is changing his focus from providing clues supposedly leading to the general area where his treasure is hidden, to telling a potential "seeker" what he (or she) must then do to actually find the treasure. Let's start with the first line in stanza #4, "IF you've been wise and found the blaze." I think Fenn could use the word WISE to mean smart, or simply aware, as in, "I'm wise as to what you are up to." As for the word BLAZE I thought it might refer to "blazing a trail," or simply providing signs of some kind or other leading to the treasure, but I somehow doubt that. Yes, I think we should look for pointed arrows and such, but I wouldn't expect to find such obvious signals. Fenn is definitely telling that "wise" searcher that has located the "blaze" to "look down." Does this mean the chest might be hidden in some dark hole, but not a cave, that one will need to look down into, to see the chest? And we must remember that the finder of the chest must be able to extract it from the hiding place. Fenn once mentioned that his chest could be found with a flashlight—could this be a clue that his treasure is hidden in some dark hole where it can only be seen with a flashlight and could the "blaze" refer to a glaring light given off by the chest when hit by a flashlight beam? That's pretty far-fetched, but possible. So the big question is—once we're down on La Madera Arroyo what can we expect to find to lead us to that treasure? I think one thing is clear, only an on-site exploration has any chance in pinpointing the treasure's location, and it might take more than a quick one-day trip.

There are some lines in Fenn's poem where I can't find much meaning at all. For example, the only interpretation I can make of "tarry scant and marvel gaze" is, "don't hang around looking at the scenery" after one has found the blaze, or sign, pointing out the treasure's location. He also tells us to go in peace. A synonym for peace is quiet. Perhaps he is telling the finder to get the chest quickly and go in peace meaning quietly, so as not to draw attention. As to the line, "look quickly down" does that mean we will find the chest only by "looking down" as from a bridge or even an embankment? He is definitely telling us to be "looking down" from somewhere. But why "quickly?" Does this mean that the finder might be in danger if he (or she) doesn't get a move on? I have never understood those words.

Another sentence I have a lot of problem with is in stanza #6, "So hear me all and listen good, your effort will be worth the cold." Northern New Mexico is not necessarily cold all year round. If his treasure is, in fact, hidden there, is he telling us it's on a tall mountain, where it would always be cold? That would destroy my ideas, but would any location on a mountain fit the description of stanza #3?

Could he be telling us that we must wade in cold water to get to his treasure? That seems to be a reasonable interpretation—take your waders along on your search.

Finally, I want to add one thing further. During my research, and after studying the maps, I keyed in "Forrest Fenn Statements" to see if I could find anything from him to support my ideas. The first site listed was entitled, "Seeker's Recap of Forrest's Statements." Now let me emphasize here what any serious "seeker" of Fenn's treasure already knows; there is just a ton of information on the internet on statements Fenn has made subsequent to the publication of *The Thrill of the Chase,* and I've read a bunch of these plus many "interpretations" of his remarks by "seekers," and I never found that "research" to be particularly rewarding. But I found the site above to be of interest. It's an actual question and answer session between Fenn and some of the "Seekers." No, I didn't see anything there in reference to my ideas, but he said something very interesting. During this session he refers a number of times to what he calls the "first clue," and states that his treasure can never be found without solving that "first clue." In fact, he says, if the searcher has not solved that, they might as well "stay home and play Canasta." That certainly got me to wondering if I have solved that all important "first clue."

He did not say that nobody has ever solved that clue. So as for my own efforts that I have outlined in this essay, I would say that IF I have solved the "first clue," it would probably be in my interpretation of "in the wood" and associating that with the La Madera Arroyo. I surely do not find any important "first clue" in his first stanza, so I assume it can be found anywhere in the poem.

So there it is, after all my study of the maps and other material, the La Madera Arroyo is the best prospect I can come up with as to the general location of Fenn's treasure. I hope anyone heading out to northern New Mexico in search of the treasure will do a lot of research before he or she goes, and will go well prepared. I'll admit I would like to go down there myself and search that arroyo. But I'm eighty years of age and a heart patient. The joke is I would go if I could get my cardiologist to go with me. But if any of you readers make the trip, I

Marvin Brooks

wish you the best of luck. I have always found Kipling's "The Explorer" to be inspirational. These following lines seem to be directed at the Fenn treasure seeker.

> "Something hidden.
> Go and find it.
> Go and look behind the ranges—
> Something lost behind the ranges.
> Lost and waiting for you.
> Go!"

I will close with this: I don't think the most valuable contribution Fenn has made in his actions is to provide some adventures to the few people who can afford to travel to northern New Mexico, Yellowstone National Park, or somewhere else in the vastness of the Rocky Mountains to search "helter-skelter" for that little chest, although obviously I would like to motivate those who can afford it to go to northern New Mexico and check out some of my ideas. Rather, I think his most valuable contribution is that his story provides a challenge to the thousands who have studied and become fascinated by his poem and treasure, who are not able to make those trips, but can try and figure out those most complicated and indecipherable clues, with the hope, however slight, that they can succeed and go get that chest. Truth is, I'm one of those dreamers, and that motivated me to study Fenn's clues and write this essay. Have I been successful in solving any of his clues? Only time will tell.

Toward Solving Forrest Fenn's Hidden Treasure Clues

Addendum

On 5/1/17. against my own better judgment, for I'm an eighty-year-old heart patient, I traveled down from my home in Colorado Springs to Santa Fe, then spent a day up north about fifty miles checking out those trails, and the canyon, that lead down to La Madera Arroyo. These are off Highway 84 adjacent to the little community of Medanales. Before going down there, I studied the Google satellite maps, but from those I couldn't determine if there is water in any of the many creek beds running into the main ravine, named La Madera Canyon. We know Fenn has said his chest is wet, and of course his poem talks of "water high," so I'm assuming the chest is hidden in water, in a creek, and I hoped to find water down those trails.

    I found the most prominent trail to be #104, (see the map), which actually is named The Apache Trail. That trail follows alongside the La Madera Canyon, which runs from way down in La Madera Arroyo up to a bridge over Hwy 84, then on a mile or two where it empties into the Chama River; at least it does when it's carrying water.

    Trail #104 has a cattle guard, then a locked gate, beside which is a smaller gate letting in hikers. A permanent sign at the entrance states the trail is maintained by a local health clinic. (See the photos). There was no one in sight around the trailhead and no parked vehicles. I don't know if it took any actual courage for me to go in there alone, but after driving all the way down from Colorado Springs, I certainly was intent on exploring some of that trail. I went down for a mile or more, sometimes on the trail, or in the La Madera Canyon, or up some dry creek bed. The entire area was as dry as a bone. There was no water anywhere. The Google satellite map indicated that the trail crossed a creekbed down about a mile. I thought I might have to wade across a creek—I even thought Fenn's treasure might be hidden up that creek. Now, it was very evident that in the past many years great floods have carved out those dry creek beds and the canyon, but on this day, I couldn't even find a stagnant pond. I spent some time in the canyon; it's wide and sandy, and, as you can see from the photos, quite impressive—but totally dry. I was very disappointed, although I enjoyed the hike, out there all alone. I was afraid I might encounter a rattlesnake or two. No, I saw no signs of life anywhere.

Later, I drove around Medanales, and tried to go down to the Chama River, to where the La Madera Canyon runs into the river, but I couldn't find a path down to it. Nor did I find anyone around the town to answer my many questions. On the Google satellite maps, that part of the river is also named La Madera Canyon and looks shallow and rough. I've even thought Fenn's hiding place might be in that river canyon, but Fenn talks about a creek, not a river, and I can't find any of his clues that supports that idea. The bottom line is: I enjoyed my trip, but was very disappointed in my finding that that area was totally dry, and without any creek with "waters high."

But what about the La Madera Arroyo on down the trail a few miles? Any such creek with water down there? It seems doubtful, because that arroyo empties into the big canyon that takes any water to the Chama River.

After returning home I put in a call to the business that maintains or sponsors the Apache Trail. I didn't expect much, but was referred to a supervisor who had some knowledge of the trail. She said that her organization made a yearly hike down the trail. She also said that she had gone down the trail in a vehicle and gotten stuck. She said the trail was actually open to vehicle travel. Really? Well, I saw chains and a lock, but I didn't try to open the gate. Maybe I missed something there and the gate is not actually locked—but that seems doubtful. However, if I go back down to that area to check out other prospects, I'll see if that gate is open, and if it is, I might drive down it as far as possible in my jeep, just to see what it looks like down there, even though, as I learned, I might be violating U.S. Forest Service policies. I had a nice conversation with this lady, and she referred me to another woman at a local U.S. Forest Service office, gave me a phone number, and said this woman knew all about the Apache Trail. So I put in a call to her. I also had a nice conversation with this lady, but I can't say that I learned much more about the Apache Trail. She did say that those trails were closed to vehicle travel except for authorized persons. She didn't give me any indication that the La Madera Arroyo was that roaring creek with "waters high." It appears that at this time, all that area is dry. There is a stock tank, but even that is off another trail.

So as far as my research, I have just about given up on the idea that Fenn's hiding place is in the area of the La Madera Canyon, or way down in the almost inaccessible La Madera Arroyo. That whole area is just too dry, and Fenn has said his chest is wet, not sometimes wet, but wet. I'll take his word for that. But I haven't given up on the idea that his line "in the wood" can be interpreted to mean, "in the La Madera," and I've learned through my map research that there are a number of canyons, and even a town, and a mountain range in northern New Mexico named La Madera. So far I haven't been able to connect Fenn's clues to any of them, but I'm still trying. For anyone interested in checking out those other La Madera named locations, I will refer you to the "Big Map" entitled "Santa Fe National Forest," and any number of topographical maps, all put out by the U.S. Forest Service. I have three of those topos covering the area of northern New Mexico that I'm currently studying. They are: El Rito, Ojo Caliente, and La Madera. I found the Ojo Caliente (Hot Eye) topo the most interesting, as it shows the La Madera Mountains and a canyon named Canoncito de La Madera. These topo maps are indexed with their locations on the "Big Map." All of these maps can be bought by calling the U.S. Forest Service at 505-438-5300. Be sure and ask them for their free pamphlet "Rio Chama: Wild & Scenic River." It's a good one, with a great map. I still haven't given up entirely on the idea that Fenn's chest might be hidden somewhere close by the area covered by this pamphlet, and anyone seriously searching for Fenn's treasure in northern New Mexico should have these maps.

Although I still believe "the wood—La Madera" interpretation has some merit and should be pursued, I have started to think more about his other clues such as "where warm waters halt." I have thought that that term simply refers to the Chama River, where warm waters halt and colder waters prevail, which is suitable for trout. But could it be, as a lot of searchers believe, where warm water from a hot springs enters a colder stream. Northern New Mexico has a number of hot springs including the world famous "Ojo Caliente Hot Springs Spa and Resort." I'm currently working on connecting that spa with Fenn's clues.

Still, I strongly believe that one will find the treasure only IF they are "in the wood," but other than the La Madera link, what could that clue mean? In the forest? There's a heck of a lot of forest in the Rocky Mountains. Could it mean a slang term for some place name ending in wood, for example a Cottonwood Creek or canyon? I haven't found any such name on the maps that fit that idea. (There is a recent movie out named "The Wood." It refers to Inglewood, California.)

What about the blaze? I think Fenn is telling us we will find the treasure only IF we first find the "blaze." I've given this idea some thought. Fenn is a pilot, and owns or did own a private plane. No doubt he is very familiar with northern New Mexico. Did he discover his "blaze" from the air? Is it such an outstanding or unique landmark that it might be prominent when viewed from the air? I'm looking for such a place on the Google maps.

What about the clue "Home of Brown?" Some searchers believe that it refers to some previous home of some actual person, probably long dead, such as Ranger Brown's home in Yellowstone National Park, or even Mollie Brown's home in Leadville, Colorado, above the Arkansas River. I tried to find some famous person with a New Mexico connection named Brown, and found nothing. I still believe that clue means home of brown trout.

I'm not going to try and decipher all of Fenn's clues in this addendum, but the internet has hundreds of interpretations of the clues. These are mind-boggling but I look in on them from time to time.

In closing, I will just add this: I believe Fenn's treasure will be found in northern New Mexico, in a stream or creek, after the finder has traveled some distance by vehicle, probably in a desolate area. I think the finder has "put in" his or her vehicle at some location below a river, maybe the Chama or even the Rio Grande, where there is brown trout. After traveling as far as the finder can by vehicle, he or she must walk for some distance, just as Fenn did when he hid his chest. The finder must have found the "blaze," possibly by studying maps, topos or Google satellite maps. The creek wherein lies the treasure is fast running, with a lot of heavy boulders, and the finder will need to wade out into a cold creek to retrieve the chest. I was just up to Blackhawk, Colorado. The road up there goes along Clear Creek. That creek is a good example of one with "heavy loads and water high." Can such a creek be found in northern New Mexico? Well, that's the creek I want to look at on the maps to see if the other Fenn clues fit. So far I haven't found it. Yes, I thought I would find it close by the La Madera Canyon that I visited.

Also, in closing, I should add that I'm looking a lot closer at some of Fenn's statements during the nine years since he published his book. In my essay I have already talked about a blog, found by entering "Forrest Fenn's Statements," where there is a recorded question/answer session between Fenn and some seekers, and where Fenn states that, unless one deciphers the "first clue," they might as well "stay home and play Canasta." But in that session he also tells us what "tools" we need to find his treasure. They are: his book, Google Earth, and a good map. He also states that a comprehensive knowledge of geography would be helpful. Well, I don't know about that geography part, but I have good maps and know how to use Google Earth. And most important, being an old retired guy, I have the time to pursue this "impossible" treasure hunt. If I find any reasonable prospects, I'll go back down to northern New Mexico and check them out, but I don't think I will be adding any more addenda or revisions to this book.

Finally, I strongly recommend that no one travel to northern New Mexico to look for Fenn's treasure without having some very good prospects based on their own extensive research, unless, of course, one has the time and money to look randomly in that vast area for that little chest. I wish the very best luck to all of my fellow searchers.

Marvin Brooks

Note

6/18/17 As I'm sure you have heard, a second Fenn treasure searcher is missing and the prospects of finding him alive are not very good. From the reports I have seen he was alone in a desolate area and may have tried to enter a swift stream. I want to make it clear that I did not engage in any risky behavior on my hike in northern New Mexico. I did not and would not climb any mountains or enter any swift streams. Further, I was in cellphone contact with my son on entering the trail, and on leaving the trail. At no time did I feel I was in danger. There are hiking guide services in Santa Fe—I talked to one of them, but didn't think I needed a guide for my very short hike, on a level trail. Now we know that Fenn went "in there" alone to hide his treasure, and this was probably in some desolate area, maybe even a dangerous one, but I seriously doubt that Fenn risked his life in hiding that chest, which should be a strong clue that we shouldn't either. Be careful—be safe.

Sources

Dobie, J. Frank, *Coronado's Children*, Vintage, 1930

Fenn, Forrest, *The Thrill of the Chase,* One Horse Land and Cattle Co., 2010

Clark, Taylor, *The California Sunday Magazine*, "The Everlasting Forrest Fenn," 7-15

Morton, Mary C., *Earth*, "On the Trail of Treasure in the Rocky Mountains," 2-15

Land of Enchantment Guide, "Rio Chama," Santa Fe, N.M., (505-629-5688)

High Desert Angler Guide Service, "Chama River," Santa Fe, N.M., (505-988-7688)

Wilderness.net, "Chama River Canyon Wilderness Area," University of Montana

Wikipedia, "Freestone Streams"

Wikipedia, "Trail Blazing"

"New Mexico Roads and Recreation Area Atlas," Benchmark Maps

"Santa Fe National Forest Map," U.S. Forest Service

"Rio Chama Wilderness and Scenic River," Pamphlet and map, U.S. Forest Service

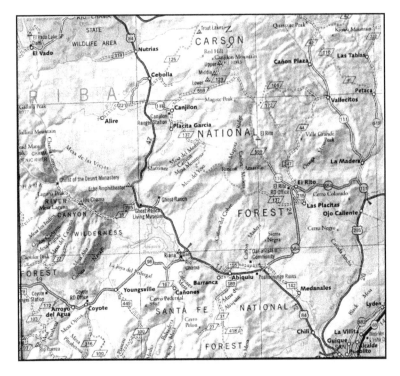

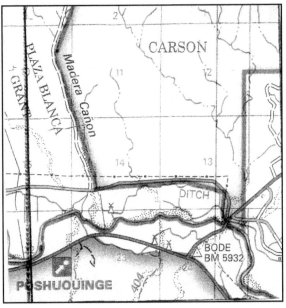

Toward Solving Forrest Fenn's Hidden Treasure Clues

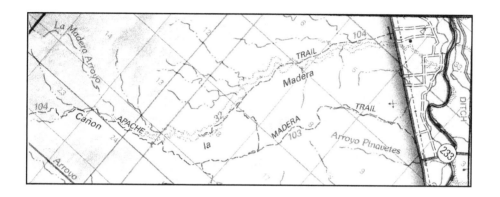

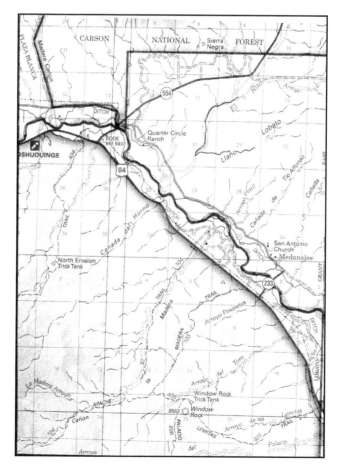

Marvin Brooks

Researching the Steamboat *Gila* Robbery– Crescent Springs Treasure Story

Chapter One

I first became interested in this story when I came across a book by Douglas McDonald in the fine little library in Laughlin, Nevada. The book was entitled *Nevada's Lost Mines and Treasures*, published in 1981, and consisted of almost one hundred short "treasure" stories. I suppose I thought of myself back then as an amateur "treasure hunter," meaning I liked to go out with my metal detector from time to time, looking for something of value. So one of the stories in McDonald's book caught my attention. It was entitled, "Crescent Springs Treasures." It covered only one page, and in summary this is the story:

A lone gunman, in 1880, pretending to be a soldier in distress, flagged down a Colorado River Steamboat (or steamer), the *Gila*, off Cottonwood Island, which I learned is now under the waters of Lake Mohave. The captain reversed engines and ordered some crew members to take a skiff over to the island to pick up the "soldier." Once aboard he pulled his gun on the crew and ordered them to load the skiff with several hundred pounds of gold and silver that the *Gila* was transporting downstream for one of the mines at Eldorado Canyon, from where the *Gila* had departed that morning, and headed off in the skiff for the western shore of the river, where he had two horses tethered to a tree.

When the *Gila* reached Hardyville, a small river crossing community several miles downstream, a posse was organized that tracked the bandit to Crescent Springs Nevada, a deserted natural spring about thirty-five miles west of the river, where they found his dead pack horse, but no trace of the booty. The posse then returned home.

Then, about 1905, after the little mining town of Crescent was founded, a miner, digging a well just north of the spring, discovered a silver bar, which was assumed to be booty from the *Gila*. There is no record of any further treasures found at or near the spring until somewhere around 1914 when a couple of boys from Crescent found a strange looking "rock" that resembled a bar of soap. Their parents recognized it was a large gold nugget, which they sold in California for enough "to move the families back to civilization." McDonald then ends his story by speculating that "a sizable cache of gold and silver could still be hidden under the sagebrush near the spring at Crescent."

Nice story, and I found it believable. And as "luck" would have it I was "snowbirding" at the time in Laughlin, a great little gambling resort on the Colorado River about ninety miles south of Las Vegas, right across the river from the larger Bull Head City, Arizona. From a topographical map I learned that Crescent Springs is just inside the Nevada border, very near the ghost town of Nipton, California, a place I had previously visited. And it was only fifty miles from Laughlin.

Winters are sunny and warm in Laughlin and sometime later, on a beautiful Sunday morning, I drove my RV over to the Crescent Springs area, intent on picking up some of that gold and silver. How difficult could it be? After all, I knew the treasures would probably be found "just north of the spring." And what an advantage I had over those folks in 1905 and 1914—I had something they didn't have—a metal detector. But of course I'm being a little silly here. Crescent Springs was only a dot on my map and there is an awful lot of desert in that area. As a matter of fact, my maps clearly pointed out that north of the spring there is nothing but desert until one reaches the Las Vegas area some sixty miles or more away. When I got over to the vicinity of Crescent Springs I could see evidence of some activity—probably some mining operations—but there was no one around that Sunday morning to answer my number one question: Where in the hell is Crescent Springs?

Still, I gave it my best shot. I found some old foundations right off the highway that I thought might be some ruins from the now-ghost town of Crescent and started my search. Seemingly I was all alone in the world out there in the Mojave Desert, with the only sign of civilization coming from a passing car about every thirty minutes. And truthfully I had an enjoyable time, for the desert is a strange place when one is alone in it. Ask anyone that has experienced it.

So I spent hours out there, until the afternoon, taking a break from time to time for snacks in my RV or to use the restroom. Now most people would not think there are any metal objects buried in the middle

Researching the Steamboat Gila Robbery–Crescent Springs Treasure Story

of the desert to set off a metal detector, but they would be wrong. As I learned there is all kinds of metal trash out there, wire, nails, bullets and shells, tin cans, and just all kinds of pieces of metal trash. My find of the day? A solitary dime, and a clad one at that. So as the day moved along into late afternoon, and with the day growing increasingly cold and windy, I reluctantly called it a day.

Well, some of us old "treasure hunters" don't give up easily, so driving back to Laughlin I decided I wanted to know more about the *Gila* robbery and the finding of those treasures at Crescent Springs. Perhaps most of all I wanted to know the precise location of the "spring;" after all, I was after treasure. Yes, I was. I figured there would be many old newspaper accounts of the robbery and the findings of the treasure in some libraries or museums in Nevada. But like many "treasure" stories, or tales, McDonald offered no sources or references at all.

A short time later, when I started to think about researching McDonald's story, I looked his name up on the internet and was very surprised to find a telephone number listed for him. Now this was all in 2012, meaning McDonald's book of 1981 was published thirty-one years ago. Would the telephone number still be current? For that matter would he still be living? Well. I placed my call and a woman answered. I told her who I was and that I was researching an old treasure story in one of Mr. McDonald's books. The woman was very helpful, although she told me she was right in the middle of a Houston traffic jam. She said she was no longer married to Mr. McDonald, but she gave me a number where I could reach him. Lady, thank you very much. So I put in a call to him and I'm glad to report that McDonald is very much alive. He answered the phone himself. Our conversation was very brief as he told me he was in a meeting. But what information was I wanting, he asked? Well, I told him, I would like to know your sources for a story on the Crescent Spring treasure from your book, *Nevada's Lost Mines and Treasures.* I could tell that he had no recollection of that particular story, and that's understandable, considering the number of stories in his book. He reminded me how long ago he had written that book and said he no longer had the files from those stories. Did he have any idea where I might find some original source? Yes, he said, you might try the Nevada Historical Society, in Reno.

Since my retirement and after buying an RV, I usually spent the spring and summer away from the Southern Nevada heat by escaping to Colorado Springs, where I have family, or near by, up in the old mining town of Cripple Creek, right behind Pikes Peak and at an altitude of almost 10,000 feet. It's now a little gambling resort and I go up there to pursue my favorite pastime, low-limit poker. But that next spring, after visiting my family and doctor in Colorado Springs, I decided to spend the summer in Reno, where I could play all the poker I wanted, and I could also visit the NHS, where I was sure I would learn all the details to "my" treasure story.

In Reno, I did in fact visit the NHS. It's in a nice building right off the campus of the University of Nevada, Reno. The research staff was very helpful, taking all the pertinent information on my "research project," and heading into the archives. They came back to me a few times for more data and I'm satisfied they did their best to find information on the *Gila* robbery and Crescent Springs treasure. But they came up empty except for a folder on the general subject of Colorado River steamboats. This folder contained a single, but very long newspaper article on that subject from *The Nevadan* dated February, 1976, by Elbert Edwards entitled, "The Day of Stage and Steamboat." It's a good article on Nevada history for the period around 1880, and while Edwards doesn't mention the *Gila* in the body of the article, there are photos of both the *Gila* and her captain, Jack Mellon. So at least I knew I was researching a real steamboat and captain, but there was no mention of any robbery. Strange.

But there is an interesting anecdote within the article about the *Gila*. It is a quote by an army officer's wife in 1874, who was accompanying her husband on his assignment upriver on the *Gila*. Edwards doesn't identify the lady or mention his source. But her comments do give us some idea as to what life was like aboard the *Gila* in that day, at least for a passenger. Here are some of the highlights from her comments: "We had staterooms, but could not remain in them long at a time, on account of the intense heat...there was no ice and therefore no fresh provisions...the fare was meagre, of course, biscuits without butter, very salted boiled beef, and some canned vegetables which were poor in those days...and that dining room was hot. The metal handles of the knives were uncomfortably warm to the touch; and even the wooded arms of the chairs felt as if they were igniting..." Well, clearly the *Gila* was not a luxury cruise ship. But the lady had high regards for Capt. Mellon: "The sand bars were a problem but Jack Mellon, the most famous pilot on the Colorado, was very skillful in steering clear of the sandbars." Little did I know then that Capt. Mellon would be a primary figure in my research of the *Gila* robbery, which was to last for the next two years, off and on of course. My poker playing took the majority of my time.

Of course I was disappointed in not finding more information, but before leaving the NHS building the staff informed me that the major newspaper in Southern Nevada circa 1880 was the *Pioche Record*, a weekly, and that they had it on microfilm. Would I like to review some of them? You bet.

At that time I had no idea as to the month of the *Gila* robbery, so I skimmed through the front pages of all fifty-two weeks, figuring that a steamboat robbery on the Colorado River would be big news in 1880. But I found nothing at all concerning the *Gila* or Crescent Springs. Hmm. That's strange. I thought sure I would find something in the NHS archives to authenticate McDonald's account of the robbery. Could he have made up that story? That didn't seem possible—too much detail-- that story had to have originated somewhere else. But where?

As I have very well pointed out, during my four months in Reno that summer I spent most of my time at the poker tables, but I did visit the Reno libraries from time to time. They have a fine collection on Nevada history and quite a few books on treasure hunting. I found two books by the very well-known writer of treasure stories, Thomas Penfield.

In one of his books, *A Guide to Treasure in Arizona*, (1973) Penfield relates a version of the *Gila* robbery and Crescent Springs treasure story. It's basically the same story as McDonald's, but he has Crescent Springs located in Arizona. Obviously this is simply a mistake—didn't the bandit escape to the western shore of the river, and don't I know the general location of Crescent Springs. Further, there is no Crescent Springs anywhere in Arizona proximate to the Colorado River, at least that I could find. But Penfield's mistake is understandable—he deals with hundreds of treasure tales. More disappointing to me was the fact that he offered no specific sources for his story.

But a second book by Penfield, *A Directory of Buried or Sunken Treasures and Lost Mines in the United States,* was much more helpful. He lists two earlier sources on the *Gila* robbery, one, an article from 1971, another from 1963. I will talk a lot about those two articles as my research continues, but at that time in Reno, I only knew that I had discovered four *Gila* robbery stories, and I was mostly intent on finding primary sources, such as newspaper accounts of the events.

Then before I left Reno I came across an old blog on the internet where a woman named Chrissy was seeking information on the *Gila* robbery story. She indicated she wanted this for the Searchlight, Nevada library. Now Searchlight is only thirteen miles west of Cottonwood Island, where the *Gila* robbery occurred, according to my sources. I thought Chrissy might work for the Searchlight library and might have some information. So I put in a call to them. The librarian that answered said she knew of Chrissy, but she didn't work for the library, nor did she know how I might reach her. The librarian said she had no information on the *Gila*, but referred me to Jane Overy, curator of the Searchlight museum.

Subsequently, I had a very interesting conversation with Ms. Overy. She told me she had had a number of inquiries about the *Gila* robbery, "but I'll tell you one thing," she said, "the robbery could not possibly have occurred where they said it did." Why? "Because the terrain is far too rough down there for horses to get in." She said she had even taken some inquirers down to the lake to prove her point. Then Ms.

Researching the Steamboat Gila Robbery–Crescent Springs Treasure Story

Overy gave me some good advice. "Focus your research on Yuma, the home port of the steamboats. If there is anything to the robbery story, you should find the answer there."

Addendum

I've since learned that Elbert Edward's account of the military officer's wife telling of her trip upriver on the *Gila* is from the book *Vanished Arizona* by Martha Summerhays, and that she traveled on the *Gila* in September, 1874, when the temperatures ranged from 107 to 122 degrees.

Chapter Two

Now I'm not a complete doofus. I fully recognized that any 1880 Colorado River steamboat robbery would be recorded somewhere in the NHS archives or the major Southern Nevada newspaper of the day, and that the stories of the robbery were probably fiction, or maybe a hoax. But I thought that since I had gone this far I should check in with Yuma, and visit some libraries or museums down there. As Jane Overy advised, if there was anything at all to the *Gila* robbery story I should find accounts of it down in Yuma, home port of the Colorado River steamboats. I envisioned myself down there poking around among dusty and yellowed newspapers of 1880, stashed in some dark corner of some museum. And I really hated the idea of going all the way to Yuma for what would probably be a "wild goose chase." Still, I wanted to be sure I wasn't missing something before declaring the McDonald and Penfield stories a bunch of, well you know what.

How utterly naïve I am. How totally out of touch with the modern world I am. I didn't need to go down to Yuma. With a little internet research I learned that the major newspaper in Yuma in 1880 was the *Arizona Sentinel,* and thanks to the Arizona Memory Project, a division of the Arizona State Library, it is all digitized (whatever that means) and available on line, yes, all the way back to 1880 and beyond. Furthermore, the researcher can designate their subject of interest, which will then be highlighted on any page under review. Fantastic! The staff at the library very patiently instructed me by phone on how to use the "Project" system, and I was on my own.

Well, the *Sentinel* proved to be a weekly, and I spent two or three days reviewing the year 1880, and a few years later. I highlighted every subject I could think of pertaining to the *Gila* robbery, or the Crescent Springs treasure. I got hits on Gila County, Gila River, Gila City, and yes, Gila Monster. I even got some hits on the Steamer *Gila*, but these were simply little items about some particular couple booking passage on the *Gila* to go upstream to visit a relative, or news that the *Gila* took on some particular cargo. There were also some items on Capt. Mellon, usually just noting that he was visiting Yuma. It appears the good captain was a well-known person in Yuma. But there was nothing

Researching the Steamboat Gila Robbery–Crescent Springs Treasure Story

about any steamboat robbery; nothing at all about Crescent Springs, or about any Hardyville posse.

That was it. I had gone far enough in trying to authenticate this obvious piece of fiction. Now it was time to move on. Where? Well, I wouldn't say I was now obsessed with the story but I was pretty well hooked on it. I wanted to know when and where the story, obviously a tale, originated, and by whom. It was during this period that I made some very interesting contacts, albeit very brief ones, with three exceedingly talented and accomplished individuals. All three are experts and writers on the history and folklore of the Old West. Why did I contact them? Basically, I wanted to know if they were aware of the *Gila* robbery story, and if they were, what did they make of it.

My first brief contact was with Bob Boze Bell, editor of *True West* magazine. While surfing around on the internet, trying to find information on Colorado River steamboats, I came across one of his websites in which he indicated an interest in that subject. I didn't know anything about Bob at that time but I learned he is an excellent artist who specializes in themes of the Old West. He even had a fine drawing of a river steamboat, mid-stream, on his website, which could very well be the *Gila*. Additionally, I learned that Bob is the author of numerous books and countless articles on the history and legends of the Old West. Bob also referenced a book on his website, *Steamboats on the Colorado River*, by Richard E. Lingenfelter, published in 1978. Now I figured that it would be a real long shot getting through to Bob, as he must be among the busiest people on the planet, but I put in a call anyway. To my surprise he took my call and listened patiently as I explained that I was trying to authenticate a well-published story about an 1880 robbery of a Colorado River steamboat, the *Gila*. Then I asked if he was familiar with the story.

Bob said he wasn't, but said he knew an excellent researcher who might find something on it. I learned a short time later that Bob set up a website, which remained on the internet until recently, entitled, *Calling All Researchers This Means You Gay*. The researcher is Gay Mathis and he must be pretty good. Almost immediately he found yet another version of the *Gila* robbery story, this one from Phillip I. Earl, who is, as I later learned, a very respected historian on Nevada folklore and the retired Curator Emeritus of the Nevada Historical Society. The article in question was published in 1991 in a newspaper, the *Parumph Mirror*. I would contact both Earl and Richard Lingenfelter in due course. But before discussing those contacts I need to mention that the website set up by Bob Boze Bell was not real successful. Part of that might have been in the name. When one keyed in the title, the website was listed alright, but the internet, in its infinite wisdom, asked the question, "Do you mean, Calling all researchers this means you're gay?" Bob mentioned me by name on the website so when I tried to impress my friends by telling them about it, they told me they tried to find it, but only found a bunch of stuff about gays. Nevertheless, I appreciated the efforts by Bob and Gay Mathis.

My second contact was more indirect. I was able to find a copy of Richard Lingenfelter's book and if there had been any Colorado River steamboat robbery, I figured he would surely mention it. His book is a fine history of the Colorado River steamboat era, with many illustrations, including photos of the *Gila* and Capt. Mellon. But of course there is no mention of any steamboat robbery. Now I learned that Lingenfelter is, in fact, a scientist, specifically an astrophysicist. Although he has written several books and many articles on old west history, that is only his avocation. He has written many more books and articles of a scientific nature, with titles I can't even pronounce.

So I found an email address at the university where he conducts research and sent him a note. Basically, I just told him of my research; told him some reason why I doubted the story of the *Gila* robbery; asked him if he had seen the story, and if he had, what did he make of it? Early the very next morning I had his response: "Thanks for your note. I've seen the story and it's entertaining—but I found no evidence that it's true. I'd say it's pure fiction. Best wishes, Rick." When in doubt about anything, it's nice to have imput from an expert. Of course the *Gila* story is fiction. Now I only needed to find out who the "culprit" was that perpetrated this hoax, and how he (or she) managed to interject so much detail—some based on truth—into the story. That would all be much more difficult than I thought. The "culprit" turned out to be pretty clever, pretty clever indeed.

My third contact was made after I tracked down Phillip Earl. Sometimes I think I should have been a private detective. I learned by contacting the NHS that he still visits there from time to time, and through them I was able to reach him by phone. He listened patiently while I told him about my research project and reminded him that he had written an article on the very subject of my reseach, which was published in the *Parumph Mirror,* a newspaper, in 1991. I told him that his article did not include sources and I was particularly interested in those as I was trying to determine the origin of the story. Mr. Earl said he did not have any recollection of that particular article, but told me he had sent out a series of articles back then on Nevada folklore to various Nevada newspapers. He explained that these type stories, folklore, written for entertainment and not serious history studies, do not usually require a listing of sources. I very much enjoyed my conversation with Mr. Earl. It was obvious that he had at least two passions—Nevada history, and writing, and he is a great storyteller. And I recognize that he is absolutely right. Treasure stories are just a form of folklore, and I have read literally hundreds of them. No one should confuse such tales with serious history subjects. Few offer any sources at all, and virtually none offer reliable original sources. But here is the kicker: eventually I would have NINE published *Gila* robbery stories in my possession. One might think that in all those stories some author would question whether or not the story was true or wonder about the original source. That did not happen. As we will see, they were all published as true, including the original. But I'm getting ahead of my own story.

Thanks to Penfield's *Directory*, I knew that, to my knowledge, the two earliest published versions of the *Gila* robbery story were Joseph

Marvin Brooks

L. Simas' "Lost Gold of the Colorado River Pirates" in *True Treasure*, June, 1971, and the earliest of them all, Fred L. Kuller's "Piracy on the Colorado," published in *Frontier Times*, March, 1965. I then set out to locate copies of these two magazines. But by this time I had reached two conclusions. One, that there was no credible evidence to establish that there was any robbery of the steamboat *Gila* in 1880, or any other time, and two, that there was no evidence of any published account of such a robbery from 1880 until the 1965 article in *Frontier Times* magazine, meaning the story originated in that article. Well, I might have been right, but those conclusions turned out to be pretty complicated, pretty complicated indeed, as you the reader will soon see.

Chapter Three

I desperately wanted to find copies of those two early versions of the *Gila* robbery, in particular the 1965 version. But that was in a *Frontier Times* magazine published forty-seven years ago, and they had been out of business for a very long time. I spent considerable time contacting a number of treasure magazine dealers operating on line without any luck, and even contacted the Frontier Times Museum in Banderas, Texas, seeking a copy. No one had a copy. But finally I learned about Ted Doades of Dundee, Oregon, who advertises nationally in treasure magazines. The editor of one of these magazines called Ted a treasure magazine dealer extraordinaire. A short time after contacting him, I had my two magazines. I think I should add that, although I've never met Ted in person I have had numerous telephone conversations with him and consider him a friend. Ted is a natural storyteller, and has kept me entertained with his treasure tales many times. When it comes to treasure hunting Ted is a true believer who has been at this pastime for many years. Now about the magazines. They arrived in good shape. Yes, they were showing a little age, but they were both intact. The 1971 article was unremarkable except it had a very good hand-drawn map showing the exact location of Crescent Springs. Otherwise, it was just another version of the *Gila* robbery, much like the others that I had on hand. And I now had the long sought earliest version in my possession, or at least I thought.

I wish I could simply reproduce a copy of Fred L. Kuller's article in full for you readers, but since that is not possible, I will have to do a lot of summarizing. I like to think about the article as being in three parts. Now, in part one, the introduction, Kuller starts by telling that, in the middle of June, 1880, a dusty soldier is waiting "anxiously" on the upper end of Cottonwood Island, pacing back and forth, checking the west side of the river to see if his two horses are still tethered among the trees. He checks to be sure his Henry rifle is fully loaded (this lever action rifle, still sold today, was made famous by the old western movies. Think of John Wayne's rifle).

Now how this bandit got out to an island in the middle of the Colorado River, with his rifle, we do not know. Nor do we know how Kuller knew what the outlaw was doing. But Kuller tells us more; he tells us about what occurred that very morning about twenty miles upriver at Eldorado Canyon, home of the mines. Capt. Jack Mellon "had directed the loading of half a ton (that would be a thousand pounds) of silver bullion onto the deck of his steamer, the *Gila*, and had personally placed a strongbox containing nearly 300 ounces of gold into the small safe in his cabin." Kuller also tells us that the proceedings were carefully supervised by a "primly" dressed Philadelphian, Wharton Barker, Secretary–Treasurer of the Southwestern Mining Company. There will be much more about Barker later in the story. Then when the bullion was loaded, "the *Gila* gave a parting blast on its whistle and churned out into the swiftly flowing current." Interesting writing. In third person, of course. I don't believe Kuller was there. Nor does he offer any explanation as to how he knew anything about what happened that morning in Eldorado Canyon, in 1880. In fact Kuller's article is without references or sources. And he goes on with his story, which I'm summarizing.

    Now this story really gets weird. Shortly before noon, the *Gila* rounds the bend above Cottonwood Island. Kuller tells us that an eyewitness to the ensuing robbery, taking the steamer down to Hardyville, "later related the story to the editor of the newspaper published at nearby Ivanpah." This account was reprinted in the *San Bernardino Times* of June 24, 1880, under the heading: "Piracy on the Colorado." The entire *Gila* robbery story is then told as a direct quote. Kuller claims he got this story from the *San Bernardino Times*, which was, supposedly, a dispatch from the editor of the Ivanpah newspaper who in turn was relating an eyewitness account of the *Gila* robbery. WOW! Now you readers can see where my conclusion that Kuller's 1965 article was the earliest account of the *Gila* robbery might be in doubt. By any means, before going into the "eyewitness" account of the robbery, which I consider Part Two of Kuller's article, I would like to tell you of my research into Kuller's claim of finding that article about the *Gila* robbery in the June 24, 1880 edition of the *San Bernardino Times*.

Researching the Steamboat Gila Robbery–Crescent Springs Treasure Story

First, was there a newspaper in Ivanpah (that's in California, in the Mojave Desert, and very close to the Nevada border)? Apparently there was. Somehow I learned that the *Journal of the West* carried two articles about the Ivanpah newspaper in business in June, 1880, one from October, 1962, the other July, 1964, both from the same author, Karl Shutka. I searched the internet diligently for other sources on this subject, and although there is a short history of the Ivanpah area on the internet, it contains nothing more than the information in the Journal. Therefore, everything that follows came from those two articles.

The newspaper was named the *Green-eyed Monster*, after a mine of the same name in Ivanpah. It was probably the most short-lived newspaper of all time, as it lasted only three months. Its editor was named Wilmont D. Frazee, but the town folks called him "Humbug Bill." That nickname came about because, in his ads, he was always saying, "this is no humbug." Now it didn't take Frazee long to realize that there was very little news in that small mining town. It appears he was some kind of Mark Twain wannabe, for in the very short period he was in business he started sending out "dispatches" to various newspapers, the *San Bernardino Times* being one. I'll try to summarize a couple for you.

Perhaps his best story is about Canteen Fish. Frazee describes these little fish in great detail, and goes on and on. But the gist of the tale is this: when the wells in the desert commence to dry up in the summer the little fish bloat up with water and strike out in schools to the next nearest source of water. Apparently, white men never see these fish, but the Indians do. Another of his tales has to do with how hot it is out on the desert. Lizards get so hot they fill up with steam and become airborne. When the sun goes down they cool off and drop back to earth. Some of these lizards have been found as far away as Prescott. Those are pretty silly tales and Frazee might well have been capable of concocting a tale about a Colorado River steamboat robbery, and sending it off as a "dispatch" to the *San Bernardino Times*, just as Kuller tells us. There is just one big problem—he could not possibly have written the *Gila* robbery story, and I have conclusive proof from more than one source.

Marvin Brooks

But before getting into all that, I want to add a few things about Karl Shutka, and his two stories.

I've already mentioned that I couldn't find anything about the *Green-eyed Monster* other than Shutka's articles, but what about Karl Shutka himself? There is mention of him on the internet, but, to my frustration, nothing is dated. However, reference is made to the "non-profit" Aranar Institute, which titles Shutka a "Senior Fellow." Apparently this organization had a ship, possibly named the *Aranargo*, and there is a good photo of Shutka in the ship's library. When was the photo made? Who knows? But Karl appears to be of a rather advanced age. There is a nice quote by the company president: "Avanar seminars and excursions restore our bodies, minds, and spirits." Sounds like a nice gig for Mr. Shutka—it appears he was an historian—but I was unable to find out anything more on him, including any published books. There is reference to an email address for him, but it is no longer in service. Now just a couple of things more. Shutka's articles contain photos of "Humbug Bill" and a frontispiece for the *Green-eyed Monster* dated Saturday, May, 15, 1880. Vol. I no. 3. Under miscellaneous ads, there is a store with an ad stating: SELLING OUT! SELLING OUT! NO HUMBUG!! Finally, I want to point out that there is one strange connection between "Humbug Bill's" actually writings, as reported by Shutka, and the article supposedly written by him telling of the *Gila* robbery. You'll see.

So where are we now in this "research project?" Well, I have found a number of articles on the *Gila* robbery, and I will find more. They have been published without questioning the story's authenticity, which demonstrates a lack of effort toward serious research. I've finally located what I thought was the original source of the story, a 1965 article in a popular magazine, only to find the author claiming the story originated from a June 24, 1880 article in the *San Bernardino Times*, which was sent by "dispatch" from the editor of a newspaper in Ivanpah, California and is, supposedly, a direct quote from an eyewitness to the crime. But we already know that the introduction to the 1965 article is pure fiction since the author could not possibly have know of the events he tells us about, such as the actions and thoughts of the bandit out there alone on Cottonwood Island, Henry rifle at the ready. But if nothing else, I'm a curious guy, and I wanted to know if there was such an article in the June 24, 1880 edition of the *San Bernardino Times*, even if it proved to be a piece of fiction by "Humbug" Bill Frazee, or if the whole story was made up by author Fred Kuller?

With a little research I learned that the John M. Plau Library in San Bernardino holds the *San Bernardino Times* for a period including 1880. I contacted their Research Coordinator, Brent Singleton, and asked if he or a staff member would review the *Times* for June 24, 1880 to see if there was any dispatch from Ivanpah on that date and in particular any article entitled, "Piracy on the Colorado." Mr. Singleton responded by email indicating there was no dispatch or any such article on that date and even sent me a copy by attachment of the entire newspaper of that date. He also sent me a dispatch from Ivanpah of June 26, 1880. News about the *Gila* robbery? No, not at all; it was about routine mining activity.

This information from San Bernardino was pretty compelling evidence that Kuller's version of the *Gila* robbery did not originate from an article by Frazee in the *San Bernardino Times*, but there is even more solid evidence that Kuller's article is fiction, in the article itself. Now I doubt if Kuller ever intended his article to be taken for anything but a tall tale, written for entertainment. I could readily see that the *Frontier Times* magazine specialized in tall tales. If alive, Kuller would no doubt be amazed that his story has been retold so many times over an almost fifty-year period to the point where it may be on the way to becoming a legendary tale of the Old West. But I had a serious question. Why would Kuller place a particular individual on board the *Gila* during the robbery who could not possibly be there? This is very strange, as you will see in the next chapter.

Chapter Four

The eyewitness in question is identified as Wm. Balderson, the deputy recorder of the Eldorado Mining District, who, according to Kuller, was taking the *Gila* from Eldorado Canyon down to Hardyville. Now how or why he ended up in Ivanpah telling his story of the robbery to the editor of the *Green-eyed Monster*, who must have been "Humbug Bill" Frazee, is not revealed to us. But, again according to Kuller, this is his story, as told to Frazee, who sends the story in to the *San Bernardino Times*, that prints the story as a direct quote from "the editor" of the Ivanpah newspaper, which is "re-printed" in its entirety and in quotation marks by Fred L. Kuller in his 1965 story in the *Frontier Times* magazine. So in effect, what follows is Balderson's story as told to Frazee, who apparently paraphrases Balderson.

He tells us in great detail how the bandit got aboard the *Gila*, pointed his Henry rifle at Capt. Mellon, ordered the safe to be opened and ordered the skiff to be loaded with the gold and silver. Up to this poine the story is fairly believable, but the story continues:

> About this time Mr. Wharton Barker, treasurer of the mining company, who has been in his own cabin, entered the Capt.'s cabin unaware that an act of piracy was in progress. The soldier ordered Mr. Barker to surrender his arms, which he instantly did, producing a small British bulldog pistol. He then demanded that he turn over the key to the strong box, but Mr. Barker denied having such.
>
> The soldier was somewhat angered by this and concluded to take a fancy to Mr. Barker's "gay and festive" suit of clothes, demanding that Mr. Barker exchange them for his own modest duds. Mr. Barker maintained that this was outrageous, but the bold pirate insisted, giving him a short lecture upon the Henry rifle. Whereupon the treasurer generously "volunteered." The soldier's garments were of a much leaner cut than those of the treasurer and though the soldier looked quite handsome in his end of the "bargain," the treasurer being a rather stout man, had considerable difficulty "making both ends meet" as it were. Mr. Balderson roars with laughter every time he tells

of the distinguished treasurer stuffed like a "sassinger" into that dusty, tight fitting soldier suit.

Have you ever tried changing clothes with one hand while holding a rifle on a boat load of tough river men who want nothing more than to kill you. I haven't either, but I sure wouldn't want to try it. I suppose if the bandit had any trouble, such as getting the clothes off and on over his boots, he would simply say, "would one of you gentlemen please hold my rifle for a moment?" The whole story is obviously a farce, but here's what hacks me off—not one of the eight other retellings of this outrageous tale mention the exchanging of the clothes. They retell the story as being true to unsuspecting readers, knowing full well that no one would believe it if they mentioned that part of the farce. And this is dishonest. But having said all that there are still some strange and mysterious aspects of Kuller's tale that must be investigated, at least by this curious researcher.

For example, who was Wharton Barker, was he a real person. Yes, he was—very much so. Kuller had it partly right, he was definitely a Philadelphian, in fact a very prominent one. And it is true that he was, as a member of the Barker Bros. Banking firm, a one time absentee owner of the Southwest Mining Company, and when it was later owned by his uncle and namesake, Joseph Wharton, he served as treasurer of that company. I should add that Joseph Wharton was a steel and iron tycoon, and founder of the Wharton School of Business, part of the University of Pennsylvania. (As I write this, a person very much in the news, one Donald Trump, brags of being a graduate of that school). But after some diligent research, I never found any evidence that Barker ever visited Eldorado Canyon. Besides, his farce aside, Kuller has the dates of Barker's and Wharton's involvement in the Southwest Mining Co. all wrong.

In fact, Wharton Barker was in Chicago in June, 1880, making history. He was at the Republican National Convention where U.S. Grant was going for a third term as president. Barker was instrumental in denying Grant the nomination, and insuring that it was given to Garfield. In a history of that convention available on the internet this is said of Barker: "His plan had succeeded. In the face of a group of determined professional politicians, Wharton Barker, a political amateur, had masterminded one of the most remarkable nominations in our history." Pretty impressive and this is the man Kuller placed on board a Colorado River steamboat, during this same period, being ridiculed because, according to Kuller's farce, Barker was forced to exchange clothes, at gunpoint, by a pirate.

Since Kuller introduced Wharton Barker in his farce I think a short summary of Barker's life and achievements are of enough interest to include in this research project. Wikipedia provides a fine biography of Barker. He commanded a company in the Civil War before age twenty; was a member of Barker Bros., a prominent banking firm; was appointed in 1878 as special financial agent to the Russian government; and founded a magazine, the *Penn Monthly*, later renamed *The American*. There is a very interesting article by Barker on the internet for anyone interested in the Civil War. The article concerns a visit Barker had with Tsar Alexander II while he was serving as financial agent to Russia. In very brief summary the Tsar apparently bragged to Barker that Russia was instrumental in winning the Civil War for the Union. Why? Because he threatened to go to war with Britain and France if those two countries entered the war on the side of the Confederacy. And to back up his threat the Tsar had sent two battle ships to New York Harbor. Of course, I didn't know any of this, but it proves to show what interesting things a person can learn while researching a simple treasure tale. But how did Barker ever get involved in Eldorado Canyon mining?

There is no biography of Barker, nor is there any comprehensive history of the mines at Eldorado Canyon that I could find. However, I was able to dig up some of the information I needed out of Don Ashbaugh's *Nevada's Turbulent Yesterday,* published in 1963. In summary, this is what I found.

Researching the Steamboat Gila Robbery–Crescent Springs Treasure Story

In 1880 a group at Eldorado Canyon built a new mill, and organized the Lincoln Mining Company. They sold out to a second group a year later, "who subsequently disposed of their holdings *seven years later* to the Barker Bros. Of Philadelphia," who then organized the South West Mining Co. At the turn of the century, the second generation Barker's overextended their operations, went bust, and the South West Mining Co. was picked up by Joseph Wharton, Barker's uncle. Although I could find no positive proof that Wharton Barker ever served as secretary-treasurer for his uncle, it is entirely possible. Strangely enough, I learned that Joseph Wharton visited his mines at Eldorado Canyon from time to time, but I found no evidence that he traveled there by steamboat. Rather, he took the railroad to Kingman, Arizona where his mine superintendent met him with horse and buggy. Let me say it again: I found no evidence that Wharton Barker ever visited Eldorado Canyon.

Wharton Barker entered politics after becoming disenchanted with the Republican Party. In 1900 he ran for president on the Populist Party ticket. He died in 1924, at age seventy-four. The steamboat *Gila* also had a long life. She was refurbished in 1899 and renamed the *Cochan*. Ultimately she was dismantled and put out of service in 1910. Capt. Jack Mellon died in 1924, at age eighty-three. In his later years he was referred to as "the ancient mariner of the Colorado," and "the old man of the river." And what about "Humbug Bill Frazee? Well, Shutka tells us that he left Ivanpah "one hot afternoon to vanish forever from the pages of desert lore—his ultimate fate only a question of what might have been."

At this point in my research I had become so involved with Kuller's article and proving that it was all a farce, a work of pure fiction, with a few real people added to the tale just to confuse some curious "treasure hunter" forty-seven years later, that I almost lost sight of the reason I got into all of this in the first place—the treasure at Crescent Springs. But before turning to the final chapter on that treasure, I need to add just a few more mysteries to this strange tale.

The very first sentence of Kuller's tale tells us that the soldier was waiting on Cottonwood Island in the middle of June, 1880. Quite by accident I found, in WorldCat an entry: *Log of steamers on the Colorado River, 1878-1880*. Only one library throughout the world holds a copy of that publication, the Arizona State Library, Archives & Public Records, at the State Library of Arizona, in Phoenix. There are fifty-four pages, hand-written, and prepared by Isaac Polhamus, and J.D. Godfrey, dates unknown. I believe these two were former steamer captains. Of course I knew there would be nothing about a robbery in those logs, but I wanted to see them anyway, at least for the dates in "the middle of June, 1880." I sent an email to the library asking them to review the logs of the *Gila* for those dates for evidence of any robbery. I received back an email: "I've attached a scan of the page from 1880 that should contain that date, but I don't see any entries. Please let me know if I can be of further assistance. Best, Libby Coyner." The attachment was for the date June 12, and only listed routine stops—nothing remarkable. The next entry, on the same page, specifically referred to the *Gila*, recording that it left with 140 tons of freight. Date: October 15, through October 19, 1880, and also mentioned nothing remarkable. Still, I thought the logs were interesting and the fact that they are available worth mentioning.

The next thing I want to mention has to do with "Humbug Bill's" tales as presented by Shutka, and the supposed *Gila* robbery story as presented by Kuller. In "Humbug Bill's" tales there are numerous references to "the boys" or "the boys at McGintey's (note the spelling). I take this to mean that in Ivanpah there was a saloon named McGintey's with a group of very loyal customers referred to, especially by "Humbug," as "the boys." Now according to Kuller's tale, the posse organized at Hardyville finally made it to Ivanpah, hard on the trail of the bandit. Then we have this passage, still supposedly from Humbug's dispatch:

Since mid-day Wednesday a wiry "hombre" passed through camp headed for Pioche, so he said and his piercing blue eyes and blazing beard suited perfectly Mr. Balderson's description of the bold pirate. Furthermore he was fitted up like a California street dandy, and the boys all took him for a montebank, but he treated the topplers at Mc Ginty's [note spelling] and thereby won the endearing support of all present, including myself.

Clearly, Kuller is protraying his tale as being the work of "Humbug," but with his claim of having Wharton Barker aboard the *Gila*, some seven or eight years before Barker arrived on the scene, that's just not possible, and beyond Kuller's story itself, there is no evidence whatsoever to authenticate any of this tall tale. Now just one other detail. As Kuller's story has it, when the *Gila* reached Hardyville, a posse was immediately organized which headed off toward Cottonwood Island; then on to Crescent Springs, and finally, Ivanpah. Well, even if the *Gila* robbery story proved believable, the posse story must be pure bull. It's hot, very hot, in that area in mid-June. It would take some time to get the horses and supplies ready for such a harsh trip. And there is no road or trail even today from Hardyville which is now taken over by Bull Head City to what is now known as Cottonwood Cove. The posse story, like so much else in Kuller's farce, is ridiculous.

Now I know that many of you readers will think that I have gone way too far in researching this old farce of a treasure story, and I might tend to agree with you. But please remember that all my efforts covered about a two year period and that I kept finding all those versions of the tale. And some of those characters I came across in my research were pretty compelling. So I enjoyed it all and I'm glad I pursued it as far as I did. But a big question remains: If the *Gila* robbery never happened, how does one explain that gold and silver found on two occasions, in 1905 and 1914, at Crescent Springs? I think I have some of the answers, but you may be surprised, there are some real mysteries lurking out there at the "Springs," thanks to a couple of compelling characters that can, along with Kuller, weave some more of those old "webs of deceit."

Chapter Five

Finding the origin of the "miner finds silver bar" story was easy. Remember, I said Kuller's article was in three parts. Part I is a third-person account of what happened at Eldorado Canyon on a day in mid-June, 1880. It details the loading of the steamer *Gila* with much silver and 300 ounces of gold, supervised by Capt. Mellon and Wharton Barker, and tells us what a bandit was doing or thinking while waiting for the *Gila* on an island in the middle of the Colorado River. All of Part I is obvious fiction. Then in Part II Kuller takes a different path in telling a story of the *Gila* robbery. He claims the story came from an eyewitness to the robbery, who somehow found his way to Ivanpah, California, a small mining camp in the middle of the Mojave Desert, where he told his story to the editor of the Ivanpah newspaper, which must have been "Humbug Bill" Frazee, who then sent the story to the *San Bernardino Times*, where Kuller, among all the people in the old west, found the story and reprinted it in quotation marks in a magazine article eighty-five years later. No, I won't rehash all the evidence that proves Part II of Kuller's story, like Part I, is pure fiction. But what about part III and the Crescent Springs treasure?

Well, in Part III this clever teller of tall tales reverts to a first-person narrative. After ending his tale of the *Gila* robbery, he goes on to say, "But the story does not quite end here—for there is one final short chapter to this lost treasure account that was told to the author nine years ago [that would be around 1956] by old Ike Allcott, long time resident of Eldorado Canyon." Kuller says that Ike, in his late nineties, told him that he recalled, shortly after the turn of the century, that a miner at Crescent Springs, "found a bar of silver while digging a shallow well a few hundred feet north of the springs." Kuller then goes on to speculate that this was only one bar from a larger cache hidden somewhere nearby, then ends by saying "a cache of silver bullion may still be buried in the heart of the desert country, somewhere not far from Crescent Springs." To which I say, yes, and I may win the lottery, if I bother to buy a ticket. But that's it, that's the basis of the silver bar treasure tale of Crescent Springs, told so many times over a period from 1965 through the present, or more than fifty years.

Researching the Steamboat Gila Robbery–Crescent Springs Treasure Story

Now who knows if Kuller ever talked to Ike Allcott. If he could make up the rest of the story he could make up this one. But Ike Allcott was a real person. One of the outstanding sources in my research was the aforementioned book, *Nevada's Turbulent Yesterday*, by Don Ashbaugh, published in 1963. In this book, Ashbaugh also tells about a conversation with Ike about lost treasure. Apparently Ike was something of a living legend of Eldorado Canyon. But Ike doesn't mention anything to Ashbaugh about the "miner finds silver bar" tale, at least that's recorded. Rather, he tells Ashbaugh about another treasure tale, this one having to do with a lost mine. Could it be that old Ike was a known "teller of tall tales?" By any means, even if Kuller was telling the truth, which is very doubtful, we have a recollection by a ninety-year-old-plus man, telling about some story he "heard about" some fifty years earlier, and given to us second hand. Of course that's all hearsay, Ike is not around to be questioned about his tale, so Ike's account as well as Kuller's is not real solid evidence that any silver bar was ever found at Crescent Springs. But, one might ask, in the 116 years since the turn of the previous century, have there been other reports of the finding of treasure at Crescent Springs? Well, I researched the internet diligently for any information to substantiate old Ike's tale, and found nothing. Nothing about any silver find, that is; now, as to the finding of gold treasure out there, well that's a different story. There is nothing in Kuller's tale about gold treasure being found at Crescent Springs, but remember, we have the "boys find gold" at the "Spring" in the 1914 story. Where did all that come from?

So Kuller makes no mention of any gold being found out there at the "Springs." Well, that's understandable because the idea that some gold was found there didn't enter the picture until 1975. That's when Richard Taylor, a resident of Las Vegas, published an article in *Lost Treasure* magazine entitled, "A Search for the Crescent Springs Treasure." Taylor tells about a trip he and his wife Leah made to Crescent Springs in search of *Gila* booty. Taylor makes it clear that he learned about the *Gila* robbery story and the finding of silver at Crescent Springs from an article by Joseph Simas, Jr. entitled. "Lost Gold of the

Colorado River Pirate," published in 1971 in *True Treasure* magazine, that I have previously mentioned. However, contrary to the title, Simas makes no mention of gold being found at the "Spring," only the silver bar. However, Taylor had knowledge of another story, one involving gold. It was from an article written in the *Las Vegas Review Journal,* in 1956, by the aforementioned Don Ashbaugh, who Taylor describes as a "Nevada ghost town expert." He probably was because he wrote a weekly column in the *Review Journal* on that subject for a number of years, as I was to learn. Now, in an article in 1956 (I have been unable to establish the exact month) on the ghost town of Crescent, Ashbaugh was somehow able to interview a former resident of that town who lived there as a boy, not too long after the turn of the century. His name was Foster McClure. In his 1975 article Taylor quoted in full an anecdote told by McClure to Ashbaugh which was taken directly from Ashbaugh's article on the ghost town of Crescent. This is McClure's story:

> One day I was cruising around aboard one of my burros, Buster, Maude, or Merry Legs, with a pal, Marr Morrison. We spotted a funny looking rock. It was about as big as two bars of soap and shaped like a railroad brake shoe. It was heavy and we knew it was ore of some kind. We took it to the Morrison house, which was closest. Mrs. Morrison suspected what it was and locked us in the bedroom so we wouldn't blab the fact all over town while she sent her daughter, Ora, to fetch our fathers. They immediately identified the ore—it was a huge gold nugget. Dad and Mr. Morrison caught the train at Nipton and took it to Los Angeles where they sold it. I can't recall now whether they got $3,200 or $32,000 for it—at my boyhood age either would have seemed like a fortune. Any way, it was enough for us to move to civilization and we did. Our visit this spring is the first time I've been back since.

Taylor goes on to speculate that the gold object found by the boys was not a nugget but a gold ingot, booty from the *Gila*. His basis for this

Researching the Steamboat Gila Robbery–Crescent Springs Treasure Story

"theory:" the mines at Eldorado Canyon were known to cast their ingots in soap-shaped molds. There is one big problem with Taylor's idea. I would later have Ashbaugh's complete article (more about all this later) and he relates that McClure's father was an assayer, so he should have known the difference between a nugget—raw ore, and an ingot. But on the other hand a gold nugget as big as two bars of soap is one big nugget, maybe as big as any found in Nevada, but I don't know for certain.

    I wanted to know more about Ashbaugh's article and McClure's great story and as luck would have it I was spending time in Las Vegas. I learned that the Lied Library, on the campus of the University of Nevada, Las Vegas, holds the collection of Ashbaugh's papers and manuscripts (I should point out here that Ashbaugh died in 1961 at age 60. His book referred to so many times in this story was published posthumously). I decided to visit my Alma Mater after a period of about forty-one years, and review the collection. I graduated from UNLV in December, 1971.

    I didn't recognize even one building, included the circular little library we were so proud of back in those years, but I was able to locate the new one, and UNLV's Special Collections. The staff brought out seven or eight large boxes of material of all sorts and I went through them for a couple of hours, hoping to find some interesting information on Foster McClure, and I did have some luck. I found an original newspaper clipping of Ashbaugh's article on the ghost town Crescent and his interview with McClure. In fact, Ashbaugh recorded a number of anecdotes from McClure about his boyhood days in that little mining community. He also related that McClure, at that time, 1956, operated a trailer park in El Centro, California. I noted that neither Ashbaugh nor McClure seemed to have any knowledge of the *Gila* robbery or the silver bar supposedly found earlier at the "Springs," as there was no mention of any speculation that his gold object might have been booty from the *Gila*. Which means that that idea came from Richard Taylor, nineteen years later. All in all, I was happy with my find as I left the library with a photocopy of Ashbaugh's article.

But later I started to wonder if there was still a McClure's Trailer Park in El Centro, and if there was, do any of the McClures still operate it. I thought if there were any McClures still around they might tell me something about their father or grandfather, especially if he had ever told them about finding that big gold object back in his boyhood in the desert—you know—the one that was sold for enough money to get his family "back to civilization." About ten minutes on the internet and I had the phone number of the McClure Trailer Park in El Centro, and gave them a call. Our conversation was brief, very brief: "Hi there, can you tell me, do the McClures still operate the trailer park?" "No sir, they sold it back in the 80's—a Mexican now owns it." "Do you know if any of the McClures are still around?" "No sir." "Well thank you very much." "You're welcome, sir."

But my focus on El Centro wasn't a total loss. I learned that the local newspaper, the *El Centro Valley News* had carried an obituary on Foster McClure dated November 19, 1982. Would his obituary tell the story that Foster was the boy that found a very big gold nugget back in his boyhood days? It seemed to me to be a possibility. Well, the Imperial Valley College holds the old copies of that local newspaper. I contacted their staff, told them the obituary I wanted and the dates and within a day or so they faxed me a copy of Foster McClure's obituary. Unfortunately it was unremarkable. McClure was born in Missouri in 1895; had been a resident of El Centro since 1930; was a paving contractor for a period; and built the first trailer park in El Centro. He was a WWI veteran, having served on the USS Pueblo, was a Mason, and was survived by his wife and several children and grandchildren. This was all rather interesting to me, but there was nothing about finding that big gold nugget. Later I did a lot of internet research. I did find a few stories about "boy finds gold" but none was about McClure, or Marr Morrison, for that matter.

Frankly, I doubted if McClure's story was true. I just thought that there would be some account of his story somewhere, since it is so compelling. But I found nothing to authenticate it. However, there was one big reason that caused me to question the story's veracity. As I mentioned before, Ashbaugh records a number of anecdotes told to him by McClure, having to do with his boyhood at Crescent, including this one:

> One day considerable wonder was occasioned when the stage failed to arrive on time, When it did come in, McClure recalls, the driver explained that he had been held up by a migrating hoard of thousands of desert tortoises. Nothing would stop them until the driver and the passengers got out and rolled a few hundred on their backs to hold the wave long enough to get the stage through.

Now even if I were inclined to believe McClure's "boy finds gold" tale, I am just not buying this "hoard of tortoises" story. Why this sounds like something from "Humbug Bill." It seems to me that McClure was simply entertaining Ashbaugh with some tall tales. And at that point, after receiving the obituary, my research, covering a period of some two years, off and on, was pretty well finished.

I'm sure my conclusions will be obvious to you readers. I think that Kuller's article is, without doubt just as Richard Lingenfelter concluded—pure fiction. In the sense that no one was victimized, I don't consider it a hoax. Rather I think it was just a tall tale concocted for entertainment purposes, and maybe to make the author a little money. But the tale is somewhat unique in that it has been rewritten and republished over a period of fifty years, without any of the authors questioning its authenticity, or the story's origin. As I have mentioned, I have collected nine copies of it, the last published in 2011, and there are probably others I don't know about. Then there is Kuller's strange insertion of real people into his tale—Barker, Frazee, and Allcott—in what is obviously a farce.

As far as the treasures, its obvious that I can't prove that McClure did not find that big nugget or that the miner did not find a silver bar out at the 'Springs." But I'll go this far—I'll bet my three-figure bank account that if they did find any treasure it was not booty from the *Gila*.

So what is the bottom line? I'm inclined to believe that there is no truth to any of these stories—that we have just been entertained by Kuller, McClure, Ike Allcott, and yes, Frazee—all tellers of tall tales, albeit good ones.

But there is still one mystery left unsolved. Who was (or is) Fred L. Kuller? I tried, but was unable to find any indication that he ever published any other magazine article and WorldCat lists no books by him. Was the name a pseudonym? I simply do not know. But I will give him credit—he knew a lot about the steamboat operations on the Colorado River; about Eldorado Canyon, Cottonwood Island, Hardyville, and Crescent Springs. He also knew about the steamboat *Gila* and her Capt. Jack Mellon. He even knew something of Wharton Barker although he misrepresented that very honorable man, by making him a figure of ridicule on the *Gila*. And not to forget, he knew about the mining town of Ivanpah and its newspaper, the *Green-eyed Monster* and its editor "Humbug Bill" Frazee. And I will admit this—I also learned a lot about all of the above, because prior to my research I knew nothing about any of it. For that alone my effort was worthwhile.

Well, even though I have concluded that it is all pure fiction, if I ever get out to Las Vegas again, I wouldn't mind taking the time to go down to Eldorado Canyon, just to see what it looks like down there. Then I might go on to Searchlight where I will turn east for thirteen miles to where the Colorado River once flowed freely. They say the lakes out west are really down now so maybe I can see Cottonwood Island, out in Lake Mohave. Then I would like to drive on over to Crescent Springs. With the map Richard Taylor provided in his article maybe I can find it next time. I'm not in real good shape nowadays but I could probably make that trip. But I doubt if I would be able to get over to Ivanpah, a ghost town now, way out there in the Mojave Desert, although I'll admit it would be neat to look for the remains of McGintey's, you know, the hangout of "the boys." I've been asked—if I go out that way again, will I take my metal detector? Probably. Who knows, maybe old Ike and McClure were telling the truth and there is gold and silver out there, just north of the "Springs," you know, under some of that sagebrush. Yes, of course.

The Nine *Gila* Robbery–Crescent Springs Treasure Stories

Earl, Phillip I. "Pirates of the Colorado River." *Parumph Mirror*, 1991

Harrison, Jr. Allen. "Eldorado Means Treasure." *Treasure*, Jan., 1992

Hooper, Vicki H. "Pirates Lost Silver on the Colorado River." *Lost Treasures*, Apr., 2011

Kuller, Fred L. "Piracy on the Colorado." *Frontier Times*, Feb–Mar., 1965

McCoy, James. "A Soldier in Distress." *Treasure Cache*, Dec., 1998

McDonald, Douglas. "Crescent Springs Treasure." N*evada's Lost Mines and Treasures*, 1981

Penfield, Thomas. "A Guide to Treasure in Arizona." *Treasure Guid*e, 1973

Simas, Joseph L. "Lost Gold of the Colorado River Pirates." *True Treasure*, June, 1971

Taylor, Richard. "A Search for the Crescent Springs Treasure." *Lost Treasure*, Dec., 1975

Other Sources

Books

Ashbaugh, Don: *Nevada's Turbulent Yesterday...a Study in Ghost Towns*, Westernlore Press, (1980)

Lingenfelter, Richard E: *Steamboats on the Colorado River, 1852-1916,* University of Arizona Press (1978)

Penfield, Thomas: *A Directory of Buried and Sunken Treasure and Lost Mines in the United States*, True Treasure Publications (1971)

Articles

Ashbaugh, Don: *The Las Vegas Review Journal*, "Crescent, Nevada—Ghost Town" (1956)

Edwards, Albert: *The Nevadan,* "The Day of Stage and Steamboat," (Feb., 1976)

El Centro Valley News, "Obituary—Foster McClure," (Nov.,1982)

Shutka, Karl: *Journal of the West,* articles on "Humbug Bill" Frazee and his newspaper, *The Green-eyed Monster*, (Oct.,1962 & Jul., 1964)

Newspapers

Arizona Sentinel, Arizona Memory Project, Arizona State Library

San Bernardino Times, (June 24, 1880)

Researching the Steamboat Gila Robbery–Crescent Springs Treasure Story

Miscellaneous

Website: *Calling All Researchers This Means You Gay*

Conversation with Jane Overy, Curator, Searchlight Museum

State Library of Arizona, *Log of Steamers of Colorado River—1878–1880*.

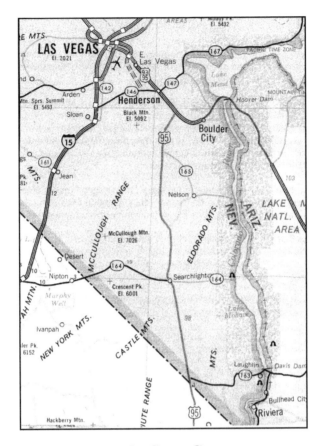

As the Story Goes

The *Gila* started out that fateful morning at Eldorado Canyon, where Hwy. 165 now ends. She traveled down to Cottonwood Island where the bandit awaited. The island is now under the waters of Lake Mohave, down where Hwy. 164 ends.

After the robbery, she traveled on down to Hardyville, now engulfed by Bullhead City, where a posse was formed.

The robber headed west until he reached Crescent Springs, just inside the Nevada border from Nipton, where he lost one of his pack animals, and may have left some gold and silver. From there he headed on out to the little mining town of Ivanpah where he bought drinks for the "boys at McGintey's," then disappeared into the night.

Researching the Missing Dimes of the Denver Mint Story

I first came across this "lost treasure" story in a library in Reno while researching the Steamboat Gila robbery: you know, that one from 1880, that I discuss in this book. I've long since lost my notes, but it was covered in one of the guides or directories by Thomas Penfield. Since I spent a lot of time in Colorado and thought of myself back then as something of a "treasure hunter," I was always interested in any "treasure" stories from that state. The entry read, "The Treasure of the Denver Mint," and added: "General Locale: In a Chasm of Gunnison River, between Crawford and Montrose, in Montrose County, Colorado." Penfield referenced a story, "New Dimes and Wagon Wheels," in a book, *Apache Jim: stories from his personal file of unfound treasures,* by Apache Jim Wilson, published in 1973. As it happened, my local library in Colorado Springs has a copy of this book, secure in its fine "Special Collections" of Colorado history and folklore. It's a small book, only 107 pages, with a number of "treasure" stories. But in summary this is Apache Jim's story of those lost Denver Mint dimes.

Jim was traveling by horseback down around those chasms between Crawford and Montrose, and was having a hard time finding a crossing of the canyons in that area. He doesn't tell us where he started his trip, but later in the story says he was headed for Pagosa Springs, although he doesn't tell us why he was headed down there. But he does tell us that his grub was running low and he needed to get some meat—snake, rabbit. deer, or elephant. Now somewhere along the trail Jim had lost his 30-30 which he blamed on his horse, Old Nig. But he still had a pistol, so secured his camp (whatever that means), staked his horses, and headed out to find that meat.

Well, as luck would have it, he spotted a "fat little doe,"got within ten feet of her, and fired. He killed the doe alright, but someone nearby let out a loud yell, thinking he was being shot at. Although Jim is not exactly clear it appears the other fellow was a deer hunter. By any means, after introductions and so forth, the two became chummy, dressed out the doe, shared the doe meat and a campfire that night, and talked about hunting and fishing. But the next day things started to get interesting.

Marvin Brooks

    The next morning after a breakfast of venison and eggs the other man had brought, Jim stared to repack his gear which included a double-box metal detector. That Jim had a metal detector excited his friend. He started to dig in his gear in his jeep, saying, "My friend, I am going to show you something that you are not going to believe!" Let's let Jim tell us what happened then: "Well, I'm standing around wondering what's going on, when he pulls out one of those old plastic-covered camera cases. He almost fell down, packing it over to where I stood. Setting it on the ground, he popped open the lid. That case was full of black dimes."

    Over a second pot of coffee, the stranger tells Jim that he had stumbled onto a place where the rocky ground had been littered with those dimes, and had spent the whole day just picking them up. I suppose he had the camera case with him, to carry all those dimes back up the chasm to his jeep. By any means, the two of them took off into the canyon with the metal detector. They found the spot where the stranger had found all those dimes, as he had marked the location with a rag tied to a stick. But they could find nary a dime, even with the metal detector, which, they soon realized, was not suitable for finding those small dimes. Not to worry, Jim had a second metal detector just for the purpose at hand. So they went back for it. After some difficulty, they found the spot where the stranger had found the dimes, searched until nightfall, and for three days more, and never found another thin dime. Well, that is almost unbelievable.

    But although Jim and the stranger did not find any more dimes, Jim did discover something of interest. He happened to see a wagon wheel in the bushes, down near the bottom of the "deep wash," and back on the rim, the two of them counted sixteen wheels. What does that all mean?

    Well, for one thing it means that the two of them spent another two days trying to get down to the bottom of the wash, without success. Then, with grub and coffee running low, they gave up the chase. The stranger told Jim a way to get over the canyons, out toward Gunnison, and the two said their goodbyes, apparently never to meet again. Well, that's just a nice lost treasure story, but where does the Denver Mint come into play?

Well, you see, on the way out of there, and before reaching Gunnison, Jim stopped at a ranch house where he spent the night. He told the rancher and his wife about the dimes, and they told him this story, which I will relate in summary: A wagon train with armed guards was shipping kegs of dimes from the Denver Mint down to Phoenix. There were four wagons carrying the kegs, and the train was accompanied by a covered wagon carrying supplies. Now the ranch couple said the men from this train had spent the night at the ranch of their parents, and left at daybreak, on their way to Phoenix. But the "old folks" claimed that a group of riders came through and told them the train had disappeared between their ranch and Montrose, where the train was scheduled for a stop. No sign was ever found of the train or the men. Well, okay, but what happened to them?

To tell the truth, we only have a "theory" put forth by Jim himself. You see the "old rancher" (that must have been the rancher who was a parent to the rancher Jim talked with) said that "during that time Indians were still roving the country. And that they were known to attack travelers and small parties." Jim believed that Indians attacked the wagon train and killed all the men with it. Then they didn't know what to do with the wagon or the kegs of dimes. Jim believed the Indians might have opened one keg, just to see what was inside, then ran all the wagons and all the dimes off the canyon rim, you know, to hide the evidence of the massacre. One thing more I need to mention. It seems these wagons, designed as they were to carry heavy kegs of coins, did not have beds, or side boards. Jim tells us that he was puzzled in seeing all those wagon wheels from the canyon rim, but no boards. Well, there is just no doubt about it, those wagon wheels, sixteen counted, were from that wagon train and the dimes found by the stranger were, no doubt, those from the keg opened up by those curious Indians.

That's not quite all the story. Jim returned to that area at a later but unspecified date, this time in a pickup camper. Would you believe, the area had changed—he couldn't see any wagon wheels or find his previous campsite. Rather, he found a new lake (I guess that would be Blue Mesa) and a lot of people. Jim ends his story by telling us how he misses the old days when he traveled through that country by horseback, you know, on Old Nig, not bounded by those pesky fences now impassable, or hindered by all those people. Lord, how Jim missed it all. But he ends his story by telling us that, "down in that God-awful chasm, there may just be a good many broken kegs with sun-blackened heaps of dimes."

Well, it's a nice lost treasure story, and a little internet research told me that people even today are intrigued with the story, and though not from around that area which is now the "Black Canyon of the Gunnison National Park," indicate they would like to visit there and search for those dimes. I'm not ridiculing those people, we treasure hunters have our dreams, and I even thought about going down there and taking a look a few years back. But today I don't think that would be a good idea, unless I just wanted to visit the Black Canyon and that whole area, now a national park. I've seen a lot of pictures of Black Canyon, and it's one scary-looking deep canyon. I have been by Blue Mesa Lake; it's beautiful, and known for its good fishing. But back to Apache Jim's story.

Look, I don't set out to debunk these "treasure" stories, or should they be called tales. I would rather be able to authenticate them, then go looking for the treasure. But once into my research, I just have to call it as I see it. One of the first things I noted about Jim's story is that there are no dates mentioned—not one. So we don't know the year that Jim was traveling down from somewhere up north to (as he tells us) Pagosa Springs, by horseback, no less. But we do know it was some time after 1941—the stranger had a jeep, and those were not produced until 1941, for use in WWII. So, if we believe the story, Jim is taking a really long trip by horseback, probably sometime in the '40's, over and in some of the most rugged country anywhere; in fact, right in the middle of the Rocky Mountains. He tells us that when he met the stranger he was

Researching the Missing Dimes of the Denver Mint Story

west of Highway 92, and after getting directions from the stranger made his way to Gunnison, where he stocked up on supplies. As I write this I have a good map of that area by my side. I suppose he took Highway 92, but it's still a good trip for Jim and old Nig. From Gunnison he tells us he "drifted" on down to Lake City, a distance of fifty-five miles. From there on to Pagosa Springs would be a real challenge, forty-two miles, including up and over Wolf Creek Pass, 10,850 feet. Of course, Jim doesn't tell us where his trip originated, only somewhere north of Black Canyon; or why he was traveling all that distance, by horseback, in the 1940's; or why he was going to Pagosa Springs. Don't ask questions, folks, just accept the story.

There were some other things in his story that caught my attention. He tells us that the stranger had filled an "old plastic-covered camera case" with dimes, and doesn't mention any other dimes at that point in his story. But when he relates details of his later trip to Black Canyon, this time in his pickup camper, he tells us about all the changes and then adds, "One thing I am dead sure of though, is my friend found around four gallons of dimes on that rim, a fact that cannot be poo-pooed away, or really too well explained." That must have been one humongeous camera case, and four gallens of dimes must have been a pretty heavy load for the stranger, climbing out of that chasm. But I'm just speculating. There is one thing more. Jim tells us he was carrying two metal detectors on his trip, a double box type and the regular type, that he and the strangers used to search for those dimes. Now I think that the double box type metal detector was developed in WWII to search out land mines, so I guess it is possible that Jim had one of those. But here is what worries me about his story, as far as the detectors are concerned. I've owned a few metal detectors in my time, and I'm even apprehensive when I leave one in view in my locked car. Yet Jim tells us about leaving at least one in his camp, while away searching for those dimes. Wouldn't he be really concerned that someone might come along while he was away and take that detector with them? I would be, but I'm just raising the question.

Marvin Brooks

There are some other "inconsistencies" in Jim's story that an old suspicious researcher like me might want to argue, but I think that's enough, at least for now. I'm sure it's clear to you readers that I seriously doubt that Jim's story is anything but a tall tale, but what about any other research? Did I make my conclusion simply on knocking holes in the tale, based on the reasoning I have presented? No, back some years ago, when I still thought of myself as a "treasure hunter," and decided to look into the story, based on the information I got out of Reno, I found a book entitled, *The Denver Mint: 100 years of gamblers, gold, and ghosts*, by Lisa Ray Turner and Kimberly Field, published in 2007. Now I felt sure if there was anything to Jim's story, it would be covered in that book. I found it listed on WorldCat and borrowed a copy. As I recall they do mention a robbery having to do with the mint in Denver, like about 1930, but there was no mention of anything having to do with Jim's story. During that time, I learned that the Mint has a website whereas people can email questions to them about the history of the Mint. I sent them details about Jim's story and asked if they could authenticate it, or if they had any knowledge of the story? They didn't respond. Now, in the present, have I done anymore research? Yes—some. I tried to find a way to contact Turner or Field to ask one of them directly if they knew of the story. Apparently they are both still actively publishing books, but I couldn't find an address or phone number for either. Then I found another book entitled, *The Denver Mint: The story of the mint from the gold rush to today,* by David Eitemiller, published in 1983. Surprisingly enough, I was able to find a phone number on David and put in a call to him. He listened patiently while I explained why I called him, and was very cordial, but he explained that his book was strictly on the history of the Denver Mint and did not get into external events such as I was asking about. Otherwise, he didn't have any knowledge of the Denver Mint's "missing dimes."

I had one more prospect for learning more about the "missing dimes" treasure story—contact the nearest treasure hunting club in the Black Canyon area and see if a member had knowledge of Apache Jim's story. I found one, the Uncompahgre Treasure Hunting Club, in

Researching the Missing Dimes of the Denver Mint Story

Montrose, Colorado. I recognized that Montrose is located on the south side of Black Canyon, but I figured any member of that club could tell me something of the local's perspective of those missing kegs of dimes, supposedly lost right there in their backyard. I found a number of a club official on the internet, and put in a call. The woman who answered was very cordial, but she didn't know much about THE STORY. However, she referred me to Mr. Detector and gave me his phone number. If anyone locally would know anything about a local treasure, she assured me, it would be Mr. Detector. So I called him.

Mr. Detector, who I learned has owned a business by the same name in Montrose for more than twenty years, is David Lehmann. From what I gather he is known by the folks of Montrose as Mr. Detector. I learned that he deals in metal detectors and coins, which I assume means scarce or rare coins. I'm glad I had an opportunity to talk with Dave. I could tell he has the curious mind of a true treasure hunter. Yes, he had heard of THE STORY. Over the years, numerous customers had asked him about it, and some even told him stories about their search for those dimes. Dave pointed out that THE STORY was mentioned, very briefly, in an atlas he had about the U.S. Mint. He read the entry to me, and except for some minor details, the entry was the same as Jim's story. Dave and I had a nice discussion about the credibility of Jim's story, and Dave said he had mentioned it once to a Black Canyon ranger who replied that it was all a bunch of bull. Dave also told me something that should be of importance to anyone thinking about making a trip to that area to search for the dimes: that whole area is now a national park and any removal of anything from the park is a violation of the law. Metal detectors are strictly forbidden, and will be confiscated—and violators may go to jail. Yes, I know, treasure hunting is becoming more difficult all the time.

Dave said he knew of a person that had a great deal of knowledge about the Black Canyon area and treasure hunting, Dell Foutz, of Grand Junction. I thanked Dave for his information—he even provided an estimate as to the weight of a gallon of dimes, based on his experiece as a coin dealer—it is about 25 pounds. That might not sound like much but if the stranger recovered four gallons of dimes; that's one hundred pounds. Wonder how he got all that weight up the ravine, and what he used as containers? But I'm digressing. I was talking about Dell Foutz. I learned immediately that he had a number of books listed on WorldCat included some of a technical nature. As I learned later, Dell is a retired geologist. But the title of one book caught my eye. It is entitled, *Elusive Treasures*, and is, I later learned, available on Amazon and the major bookstores. But as to my contact with Dell, we had a nice conversation, some of it having to do with the lack of profits for the author in the book publishing industry. But he didn't have any information on Apache Jim's story, and after I hung up, I declared to myself—that's it, that's as far as I am going on this obvious piece of fiction. I'm winding it up. But just a couple of things before I do that.

As I have already made clear, Apache Jim does not mention even one date in his story, and that alone is surprising. If his story were true, wouldn't one expect him to look at the date of the dimes found by the stranger and report that? But he didn't. Yet someone, sometime started the idea that the "missing dimes" were 1907d's, and some dimes of that date are rare and extremely valuable. Over the years there have been scores of online discussions of Jim's story. I have found no indication that anyone has authenticated the story—no newspaper accounts or anything in the history of the Denver Mint. One of the bloggers did a little research and pointed out that there were railroads in Colorado circa 1907—the Denver Mint would not be shipping coins by wagon train. And I found nothing to indicate that the local folks around Black Canyon take the story seriously. The internet accounts frequently refer to it as a legend, and that's my final conclusion. Apache Jim Wilson is an accomplished teller of tall tales, just as were many of the storytellers in my previous research project on the robbery of the Steamboat Gila. Enjoy the story for what it is, but don't take it seriously, and don't travel to Black Canyon for the express purpose of finding those dimes. But if you do, leave the metal detector at home. Mr. Detector made it clear—those federal park rangers are not to be messed with.

Now, is there anything more? Yes, one small passage in Jim's tale caught my attention. Here it is, told to us by Jim, sitting in his pickup camper, reminiscing: "I couldn't help but remember back to that time I crossed that valley with a pack string. Climbing those steep hills, southward, I had held onto old Nig's tail. It was thin and stragglylooking, but he pulled me to the top. There weren't many hairs left in old Nig's tail, but that was from pulling me up all those other countless hills back through the years." Well, I think my first thought on reading that passage was that it was too late to charge Jim with animal cruelty, but the second thought was that anyone who would get behind a horse on a steep hill, holding his tail, while being hauled up the hill, would soon find himself with about 1,200 pounds or more of horseflesh on top of him, with broken limbs on himself and the horse. You know, I have found that tellers of tall tales sometimes put utterly ridiculous passages in their tales, maybe subconsciously, that alerts the reader to the fact that it is simply a wild tale. That passage above might have been one of those warnings. I rest my case.

Sources

Wilson, Jim: *Apache Jim*, "New Dimes and Wagon Wheels," H. Glenn Carson Enterprizes (1973)

Turner, Lisa Ray and Field, Kimberly: *Denver Mint: 100 years of gamblers, gold, and ghosts*, Mapletree Pub., Co. (2007)

Eitemiller, David: Denver Mint: The story of the Mint from the gold rush to today, Jende–Hagan (1983)

Conversation with David Lehmann (Mr. Detector) Montrose, Colorado
Conversation with Dell Foutz, Author of *Elusive Treasures*. Grand Junction, Colorado

Researching the Missing Dimes of the Denver Mint Story

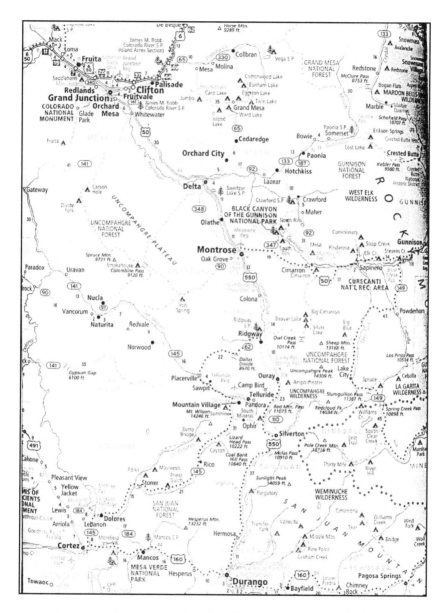

As Apache Jim tells us, after leaving the north rim of Black Canyon, he went to Gunnison for supplies. From there he traveled to Lake City, and from there headed on down to Pagosa Springs, his destination—a heck of a trip on horseback.

Researching the Missing Gold from the Steamboat *Far West* Story

This has been a favorite treasure story of mine for many years. I could have learned about the story in the book I now have before me, a borrowed copy of Choral Pepper's T*reasure Legends of the West*, which was published in 1994, or it could have been in some earlier publication. But Pepper offers a pretty good version of this famous story of lost treasure. While reading the story back then, I thought it might be true and even thought of going up to Montana to visit the site of Custer's Last Stand and look for the lost gold, based on the clues from my source. Of course, like many of my planned "treasure hunting" trips, this one never quite got off the ground. Nevertheless, the story has some historical significance, and I thought some research into the story at this time, while I am researching other famous treasure stories and writing up my findings, might be worthwhile. I think I will start with Pepper's account and go from there.

Pepper's book is not long, only ninety pages, but features eight lost treasure stories, including one on the Lost Dutchman, and another which I will also research for this book, that of Maximilian's lost treasure, perhaps the richest of all "land-based" treasures, supposedly buried out there in West Texas near a place named Castle Gap. But I'm getting ahead of myself; let's get back to Pepper's book. It contains numerous photos and original artwork. All in all, it's a fun read, and doesn't appear to be overly serious. Yet Pepper treats the stories as if they are true, and although there is not a lot of evidence of deep-seeking research, Pepper tells the stories very well. I enjoyed her book. So let's get into the one on Custer's gold. What is that all about?

As Pepper tells us, the story starts on the day of "Custer's Last Stand," June 25, 1876, when Custer and 200 of his men were killed—there were no survivors. Now at this same time a supply boat, the *Far West*, was headed upriver for a rendezvous with a General Terry, described as Custer's commanding officer. The captain of that boat, Capt. Marsh, overshot his rendezvous point and unaware that the Indians were amassing in that area, decided to tie-up for the night and head back downstream the next morning.

Now here is where it gets interesting—treasure-wise, that is. A mule-drawn wagon, according to Pepper, was "headed in the direction of the river with a consignment of gold to supplement U.S. Army payrolls at a division center in Bismarck, North Dakota." Its driver, Gil Longworth, and his two armed guards were aware of the rising Indian hostility. After meeting up with Capt. Marsh and explaining the situation, Longworth made the decision that the gold would be safer aboard the *Far West*. The Captain agreed—the gold was put on board the boat, and Longworth and crew high-tailed it back to Bozeman, Montana, where their trip originated.

Now here is where this story becomes, well, hard to believe. As night fell—this would now be June 26th—Capt. Marsh recognized he and his boat were in a precarious position, meaning there was evidence of Indians all over the place. So the captain and his two top officers, first mate Thompson, and engineer Foulk, all agreed "that the safest procedure would be to bury the gold in a suitable spot ashore until Indian unrest subsided." Now get this, according to The Story, all three officers left the boat. Who did they leave in charge of the boat? Pepper doesn't tell us. But in the dark, and who knows how far from the boat, the three men found a cave. The captain had carried a shovel and axe, and the men secreted the gold, which I suppose was in some kind of metal box, in the cave—then made it back to the boat.

Well, a couple of things. Why remove the gold from the boat? To save it from falling in the hands of Indians? Hell, if that boat were attacked by the kind of Indian force that would result in capturing it—Capt. Marsh and his two officers could kiss their "you know what" goodbye, and there would be no one with scalp intact to tell anyone where that gold was hidden, while if they made it safely, the gold would be aboard, safely delivered. But another thing, does anyone believe a boat captain would leave his command for any period of time *and* take his two top officers with him? Folks, I just doubt that. But what happened next, as reported by Pepper?

What happened was that the *Far West* made it downstream, learned of the Custer massacre, learned of another battle involving Major Reno and his troops, and engaged in transporting wounded troops to Ft. Lincoln, near Bismarck, North Dakota. Pepper tells us that it was three years before Marsh made it back up the Big Horn, at which time he learned of the fate of Longworth and his crew. As Pepper tells us they had returned to about fifty miles from the river where "their bullet-riddled bodies along side their burned-out wagon," was discovered. The Bozeman company that employed Longworth went out of business and it is assumed that they believed the gold shipment was taken by the Indians.

Well, what about Capt. Marsh and his two officers? I'm a suspicious type—did one or more of them go back and find that gold, and lived in luxury the rest of their life, and isn't there some suspicion that they took that gold off that boat, and hid it with just that evil intent? Not according to Pepper's account—"Marsh, Foulk, and Thompson, the only individuals aware of the hidden cache of gold, all passed the rest of their lives making an honest living working on riverboats. None of their future activities indicated in any manner that they had struck it rich." However, Pepper doesn't tell us how she, or anyone else, knows this. As for myself, I smell a conspiracy. But I'm being a little silly. Capt. Marsh proved to be a very honorable man. Indeed.

That's basically the story, but I have noted that Pepper mentions the *Far West* log book a number of times. Toward the end of her account she even says this: "The log book report concerning the true disposition of the gold consignment and its transfer from Longworth's wagon to the *Far West* revealed that Capt. Marsh and his two officers had completed the task of securing their charge ashore and were back aboard the boat within three-and-a-half hours." Three-and-a-half hours!! Would any ship captain leave his boat for that period of time and with his two top officers, leaving the boat in charge of who knows who? I just really doubt that. Unless there is something here that we researchers are not privy to; otherwise, the action by Capt. Marsh makes no sense, no sense at all. There is also another question. Who was the ultimate owner of that gold? Wouldn't that be the U.S. Army? Did they investigate this case? But here are the biggest questions of all. That transfer of the gold, and that absence by the *Far West* Captain—can that all be verified by the boat's log book? And where can this researcher find a copy of that

Researching the Missing Gold from the Steamboat Far West Story

log? I will attempt to answer those questions, and more. But first I want to complete my comments on Pepper's story.

Pepper tells us that she and a companion, at an unknown date, visited Hardin, Montana, which is located at the confluence of the Little Big Horn and Big Horn Rivers. They then went upriver about fifteen miles, the distance Capt. Marsh overshot the location where he was to rendezvous with Gen. Terry, and where he tied up the *Far West*; accepted the gold from Longworth, and subsequently took the gold on a trip into the hills lasting three-and-a-half hours, looking for a place to hide it. Pepper's estimate led her to the famous river crossing used by Custer, which has a sign designating the location. The land near that location is flat, so Pepper speculated that Marsh and his men traveled north to an area known as Pine Ridge, secreted the treasure, and still had time to return to the boat in the three-and-a-half-hour time frame. Now Pepper doesn't tell us how she came to be in that particular part of the country, but it apparently wasn't to engage in serious treasure hunting. She tells us: "A treasure so valuable and accessible; a site so unexploited—these factors added up to the treasure hunting opportunity of the decade." So did Pepper and her companion go in there and retrieve that so-easily accessible treasure? Well, no, you see she had Tiggy, a siamese cat, who had been left alone at home (wherever that was) and "We simply hadn't the heart to delay our return to her another minute." Well, did she return at a later date to pick up the treasure? I have no evidence that she did.

Now I assume that Pepper did not make up that story. Surely, somewhere there is her source, where I can find more information, including, hopefully, information where I can find the log book for the *Far West*, especially concerning the actions of Capt. Marsh and his crew on that fateful date, June 26, 1876. As I've pointed out, Pepper's book includes nine treasure stories, but unfortunately she lists all her sources in one bibliography at the end of the book. But the only source I found that might relate to the story at hand was *Custer Legends*, by Lawrence Frost, published in 1981. As luck would have it, my local library had a copy, and I soon had the book before me. Surely, I reasoned, this book would give me much information, and would probably be Pepper's primary source for her story. Surprise! This book does not have one mention of any lost gold connected to Custer—no mention of a Capt. Marsh, a Gil Longworth, or anything else relating to Pepper's story. Yes, there are references to the *Far West*, on five separate pages. In fact, Frost tells us of meetings on that boat between Custer and his commander, Gen. Alfred Terry. But my story is not intended to be a history of any sort as to what happened on the days preceding the Little Big Horn Battle. I simply noted that the *Far West* was an important supply boat for the troops in that area. Later, she (even steamboats are referred to as female), her Captain, the aforementioned Grant Marsh, and a horse named Comanche would all become something of heroes and I think their story is worth telling, but I'm getting ahead of myself. The facts were that I could not establish the truthfulness of Pepper's story, or anything about its origin from Frost's book, nor could I find anything in Pepper's bibliography to answer my questions. So I turned my attention to the internet, with such key words as the *Far West,* Capt. Marsh, and Custer's Gold. Fortunately, I found a website of James Deem, an author of something like twenty-five books, many for youngsters, but one entitled, *How to Hunt Buried Treasure*. Deem discusses the book on his website, and actually gives two versions of how the *Far West* came to have treasure—gold—on board, and was buried by Capt. Marsh and his crew. One of those versions was a condensed account identical to Pepper's story, and he said it came from a book by Roy Norvill, entitled, *The Treasure Seeker's Treasury*, published in 1978. My library does not have a copy, so I put in a request to borrow the book via interlibrary loan. As I write this I am awaiting the book. But my internet research led me to another potential source of information on my research "project," Dr. Douglas Scott.

Dr. Scott actually holds the title, "Battlefield Archaeologist" and his online biography told me that he had spent years studying the Little Big Horn battlefield, and many others. He has written or contributed to a least a dozen books and countless scholarly papers, and has received numerous awards. I believed if there was anything at all to Pepper's story, Dr. Scott would be aware of it, so I set out to locate a phone number or email address on him. Now as far as I can determine, I only have one talent, the ability to track down and contact well-known and accomplished people. So I located Dr. Scott at the university where he is now serving as a Visiting Research Scientist. His department has an email address where the general public can ask questions of the staff. I sent Dr. Scott a short summary of my project and asked him if he had knowledge of the story. The very next morning I had his response: "The story of the *Far West* carrying a gold shipment, then caching it is pure myth." He then goes on to say that the ship's surviving manifest showed that the *Far West* carried military supplies and equipment for the Custer expedition. He was unable to tell me the origin of the myth. And that, folks, is good enough for me—we are dealing with a myth; but I still wanted to know where Pepper got her story, and where the story first originated.

And as luck would have it I found another big lead. I keyed in the name, Gil Longworth, you know, the driver of the stage that supposedly gave his gold consignment to Capt. Marsh. I didn't expect anything, but to my surprise a hit came up where Longworth's name was highlighted in an entry on a publication entitled *Custer, the Seventh Cavalry, and the Little Big Horn: A Bibliography*, published in 2012. This turned out to be a compilation of works on the subjects of its title. It has over 2,500 works listed and I believe one will find anything ever written on those subjects. But the listing with reference to Longworth was entitled *Undiscovered: The Fascinating World of Undiscovered Places, Graves, Wrecks and Treasure* by Ian Wilson, published in 1987. Now we are getting somewhere. The book's description read, "The last chapter is entitled, "The Gold of Custer's Last Stand," with an 1898 Two Moon account of the Little Big Horn battle. The author believes "because of so many hostile Indians in June, 1876, a freight driver named Gil Longworth entrusted a shipment of gold to the steamboat *Far West* captain, Grant Marsh, who hid the gold along the west bank of the Big Horn River." Which causes me to ask, why, why would that author believe that? Well, I also had to request a loan of this book via interlibrary loan and perhaps when the two books come in, I will find

Marvin Brooks

some answers. At this point I can only wait. However, in the meantime I want to tell the story—this is a true story—about two heroes up there on the Little Big Horn, in the aftermath of Custer's and Reno's battles—a boat captain named Marsh, with his steamboat named the Far West, and a horse named Comanche. Here, in great summary, is the story.

The *Far West* was a 190-foot freighter described on the internet as "light. strong, and speedy." What's more, it only needed thirty inches of water in which to operate. As we have already seen, this boat, with her captain, Grant Marsh, was up on the Big Horn during the battles of Custer and Reno. Prior to those battles she had sometimes been a command post for Gen. Terry, Custer's commander, and it was from a meeting on board her that Custer received his last orders. For a full report on all the details, I will refer you to Joseph Hanson's *The Conquest of the Missouri,* published in 1946. But for this summary I will just point out that Capt. Marsh, at a location on the Rosebud River, was ordered to report to the confluence of the Little Big Horn and Big Horn Rivers where he was given the mission of transporting wounded troops from Reno's battle to Ft. Lincoln, 710 miles distant. Marsh took on fifty-two of the wounded and made the trip in the record time of fifty-four hours, no doubt saving some lives of the wounded soldiers. I think it can be said that he and the *Far West* earned hero status for his efforts, and Marsh was given an official "Thanks" from Congress. He remained a "steamboatman" the rest of his life, and died in 1916, at age eighty-two. The *Far West* didn't fare quite so well. She continued in service for sixteen more years, until she "hit a snag" and sank in 1883 off St. Charles, Missouri. Now, what about that horse, you know, Comanche, the only living survivor of Custer's "Last Stand."

Fortunately, my local library in Colorado Springs has a book entitled *Comanche: The horse that survived the Custer massacre*, by Anthony Amaral, published in 1961. Here is a summary of the story of Comanche's rescue, from that book. The first troops that arrived at Custer's battle site must have witnessed a terrible scene. The Indians at that time and place were known to mutilate the bodies of their victims. But the only living things at the battle site to survive were a few extremely wounded horses. Obviously the Indians took all the able-bodied ones. The decision was made to put all the horses out of their misery. However, one of the soldiers contended that there was one who could, with care, survive. That horse was Comanche and the soldier that saved him recognized him as being the mount of Capt. Miles Keogh.

The wounded soldiers from Reno's battle and Comanche had to travel about fifteen miles to the location of the *Far West*, waiting to transport them to Ft. Lincoln. Historians are not sure if Comanche walked that distance or was loaded onto a wagon, but it's generally assumed he walked. Once aboard, Amaral tells us, a bedding of grass was prepared for him near the rear paddlewheel, and a veterinarian tended Comanche on the trip to Ft. Lincoln. Comanche lived for fifteen more years, and one might say, lived the "life of Riley." You see, the Commander of Ft. Lincoln issued a General Order concerning Comanche, which read in part: "The Commanding Officer of Company 1 will see that a special and comfortable stable is fitted up for horse, and he is not to be ridden by any person whatsoever, under any circumstances, nor will he be put to any kind of work." It is said that Comanche, in his leisure, became intemperate. Lawrence Frost, in his book *Custer Legends*, tells it best: "During the year of convalescence he was given a whiskey bran mash about every other day. Convalescence soon became a pleasure. Comanche became a regular visitor at the enlisted men's canteen on pay days where the boys, willingly enough, treated their favorite to a bucket of beer. When the boys ran out of funds he would visit the officer's quarters to panhandle."

Comanche died in 1891. He had been a member of the U.S. Army since 1866, a period of twenty-five years, and was believed to be twenty-nine years old. He was only one of two horses to be given a full military funeral, the other being a horse named Black Jack, who had served as a ceremonial "missing rider horse" in over 1,000 military funerals. Comanche was stuffed and he remains on display today in a special climate-controlled encasement at the University of Kansas.

So it was that I learned a few things connected to Custer's battle—something of the life and times of Capt. Marsh, and his exploits, and his boat, the legendary *Far West*. And I truly enjoyed learning of the life and times of Comanche. Those kind of results—I think the professional researchers refer to it as serendipity—are one of the rewards of researching even old treasure tales, by a decidedly avocational researcher. But perhaps I'm getting too far off the reason for this particular project, which, if I can remember, was to authenticate or debunk the story about lost treasure, that gold supposedly entrusted to Capt. Marsh, but buried by him and his crew somewhere close to the Little Big Horn River; that treasure story told to us in Choral Pepper's book, and entitled, "Custer's Last Payroll." Regarding my research thus far, where do we stand?

Well, we've heard from a first-rate expert, Dr. Douglas Scott, who declared, without equivocation, that the story is pure myth. But we still want to know where the myth originated, and there is something else I would like to know. Was Gil Longworth an actual person—it appears that he might have been—and was he, and two guards, killed by Indians while transporting a gold shipment, and if so, did this gold disappear? And obviously, if such gold did in fact disappear, what happened to it? At the time I write this I am awaiting interlibrary loans on books by Roy Norvill and Ian Wilson, either of which might answer some of my questions, the big one being, since this story, as presented by Pepper, is simply a myth, where did the story originate? However, while waiting for the two books, I want to pursue another mystery. Where is the *Far West* log book, and what does it tell us of the events around that date, June 26, 1876? For example, does the log record that Capt. Marsh left the *Far West,* on that date, for any reason?

Pepper refers to the log book of the *Far West* a number of times in her story, but two references are of particular interest. She says, "According to the log of the *Far West*, the transfer of the gold consignment to the boat occurred 26 June." Then later, "The log book report concerning the true disposition of the gold consignment and its transfer from Longworth's wagon to the *Far West* revealed that Capt. Marsh and his two officers had completed the task of securing their charge ashore and were back aboard the boat within three-and-one-half hours." Clearly, Pepper's source, whatever that might be, relied heavily

Researching the Missing Gold from the Steamboat Far West Story

on the log book of the *Far West* for their information. So where can I find that log book? I assumed it would be in some museum some place, but even then, there must be copies, reproductions available somewhere out here. But where to look?

Well, I thought one good place to start would be in some museum or library in Bismarck, North Dakota, home port of the *Far West*. In fact, Bismarck has exactly what I was looking for—the State Historical Society of North Dakota. I immediately sent them an email, telling them of my research project, and asking their archivists if they had the *Far West* log book, or a copy, or knew where I might find such a copy? The very next day, I received an email from that institution from James A. Davis, Head of References, State Archives. Here is his opening remark:

> Thank you for your email but the *Far West* log has never been found and is probably no longer in existence. We get many requests for the log every year but it is assumed that the log was lost when the *Far West* sunk in 1883. The log is the record of the steamboat and should have remained with the boat.

Davis also makes reference to some other aspects of Pepper's story, and references the book, *The Conquest of the Missouri*. Davis tells us that the author of that book interviewed Capt. Marsh; that there was no mention anywhere of any gold shipment; and as to the title of Pepper's book, the troops had just been paid. Davis also provided a link where one can actually read the *Bismarck Tribune* for the time period in question, and further adds this: "There is no mention of gold or things of that nature connected to the expedition."

Perhaps some of you readers think I might be delving a little too deeply into this obvious piece of fiction. But I wanted to see those log book entries, if they were available, just out of curiosity, and I'm really glad that I touched base with James Davis. When I finish with this book, I'm looking forward to reviewing the *Bismarck Tribune* for late June and July, 1876. What stories it must tell. But for now I think there is only one unanswered question for this research project. Where did this story, told in so much detail by Pepper, originate? That might prove to be a difficult question to answer.

The first book to arrive via interlibrary loan was Ian Wilson's *Undiscovered*. As I pointed out earlier, this was the only book in an extensive bibliography, with about 2,500 entries concerning all aspects of Custer and the Seventh Cavalry, with a mention of Gil Longworth, the stagecoach driver that entrusted his gold shipment to Capt. Marsh, or so the story goes. But I was disappointed in Wilson's story on our subject. It was only one of thirty stories of "undiscovered places, graves, wrecks and treasures," and the Custer story was the only one located in the U.S. His Custer story, published in 1987, was only five pages and offered very little that wasn't in Pepper's version of 1994. About the only difference I could see between the two versions is that Wilson tells us Capt. Marsh ordered deck hands to row two boats ashore, to transport the gold, Capt. Marsh, and two officers. And Wilson also reports that they were back on board in three-and-one-half hours.

Wilson does mention a name I had seen before in my research, Emile C. Schurmacher, who Wilson says is an American writer that claimed to learn of "the gold story" accidentally while researching the operations of Custer's commander, Gen. Terry. But Wilson doesn't mention the title of the source, or how much of his information he got from Schurmacher. However, I assume that source is Schurmacher's *Lost Treasures and How to Find Them*. I would like to see a copy of that book. Perhaps it would give details to help determine where our story originated, but would you believe, the only copy of that book listed on WorldCat, in this entire world, is held by a library in Malaysia, and according to WorldCat, 9,000 miles away. Somehow I doubt if I would get an interlibrary loan on what must be a very rare book. So now I can only await the arrival of Roy Norvill's book, and I will continue at that time.

When delving into old treasure stories, legends and myths, the researcher is very often disappointed on finally receiving some source, usually some book, that required some special effort to acquire. So it was with Roy Norvill's book, *The Treasure Seekers Treasury*. It took

Researching the Missing Gold from the Steamboat Far West Story

me several weeks to receive the book from an interlibrary loan; in fact it came all the way from the Boston Library. As it turns out, Norvill is a British author, and although his account of our story—he calls it the Big Horn Gold—is well written, and offers a lot of detail, such as conversations between Marsh and Longworth that no one could know, there is nothing new. In lieu of a bibliography he offers a "list for further reading" for his seventeen treasure stories, none which I could relate to the "Big Horn Gold." However, in his story Norvill, although not explicitly saying his account comes directly from Schurmacher's book, seems to imply that book was the source for his information. And, to my great surprise, Amazon has copies of that book available. It's a paperback and cheap, so I have an order in. Perhaps that book, published in 1968, will prove to be the origin of this story.

    It took Amazon only a few days to get Schurmacher's book to me. Now, if the books by Norvill and Wilson proved to be disappointments, the same can be said in spades for Schurmacher's *Lost Treasures and How to Find Them*. To begin with he tells an entirely different story from what I got from my other sources. There is no Gil Longworth or stagecoach in his version of the actions of Capt. Marsh and the *Far West* up on the Big Horn River at the time of Custer's battle. According to this version, the Far West had routinely picked up a gold shipment, said to be valued at $365,000, at a place named Williston, to be delivered to Bismarck. But, as in the other version, he decided to hide the gold ashore and return for it at a more peaceful time, and then assumed the duty of transporting wounded soldiers to Ft. Lincoln. Schurmacher provides a much more dramatic version of that trip of the *Far West*, telling of an attack by Indians from the shore, who shot flaming arrows onto the boat. But the boat withstood the attack and made it on to Bismarck, to deliver the wounded troops and news of the massacre at Little Big Horn to the world and the widows of the fallen troops.

    Contrary to reports by Wilson and Norvil, and alluded to by Pepper, Schurmacher makes no mention of any *Far West* log book, and in fact offers no sources whatsoever for his story. Rather, his story is told in the third person, and is what I call "historical fiction." I ran into this type fiction in researching a previous book on the legend of Baby Doe Tabor, as well as many other stories I have researched. The authors of these type writings are not historians under any definition; rather they are writing for entertainment purposes. Their mode of operation is to take some historical event, throw in a few actual facts to lend credibility to the story, then allow their imagination to soar, as they relate events and conversations that are pure fiction. The problem for the researcher is that these writings are usually presented as factual accounts about things, such as private conversations, that the author could not possibly know. Schurmacher threw in some of the well-known facts about Capt. Marsh, and the *Far West,* and then, just like author Fred Kuller on the Steamboat Gila robbery tale and many others, used his imagination to create an entertaining story. But what a lot of these tales have in common are fictional events that simply do not make sense, such as the exchanging of clothes scene during the Gila robbery, or Capt. Marsh's decision to bury the gold entrusted to him on the *Far West*. That didn't make any sense, and these unbelievable details many times reveal the stories as fiction, myths, legends, and sometimes hoaxes.

So I went back and reread Norvill's story. Here is one of his passages regarding Schurmacher: "While investigating a missing shipment of gold, Schurmacher was researching an official operations report written by General Alfred H. Terry, which in turn led to the examination of long-forgotten boat logs." Then, according to Norvill, Schurmacher's research led him to the scene of "Custer's Last Stand." Now, even if this were all true, how was a conclusion reached that a missing shipment of gold ended up on the *Far West*? But here's the real kicker, Schurmacher's book mentions none of this—not at all, and I wonder if Norvill ever saw that book. And here is something interesting. In Norvill's book, Marsh's officers were his first mate, Ben Thompson, and his engineer Foulk. In Schurmacher's version his officers were Dave Campbell, his pilot-mate, and Ned Jenks, his bo's'n. Those "Historical fiction" writers really like to throw in those names, it lends credibility to their fiction. So with those little inconsistencies, I think I will conclude my research into the works of Pepper, Norvill, Schurmacher's and who

Researching the Missing Gold from the Steamboat Far West Story

knows how many others, all works of fiction, even though all were presented as true stories.

Based on my research, it is well established that the *Far West* was a supply boat serving the Seventh Cavalry at the time of Custer's and Reno's battles. Her skipper, Capt. Marsh, was ordered by Gen. Terry to transport fifty-two wounded soldiers from the Reno battle to Ft. Lincoln. Also on board was Comanche, a badly wounded horse of one of Custer's officers. Marsh made the trip to Ft. Lincoln in record time, and was commended by Congress for his efforts. There is no substantial evidence that Marsh had any shipment of gold on board during this period or ever buried any such gold. The accounts of this myth by Pepper, writing in 1994; by Wilson, in 1987, and Norvill, in 1978, are basically the same story, and I can find no earlier version of the story than that of Norvill. He refers to Schurmacher's story published in 1968, but that story does not contain any of the information claimed by Norvill. I found no other original versions of this tale. It appears to me that the story originated with Norvill's version, published in 1978.

Also, I was unable to find any sources to verify that there was a Gil Longworth and that he and his two guards were killed by Indians, near the site of Custer's battle, during that same period. I assume all of that is also fiction, probably by Norvill. However, if one wants to do further research, the *Bismarck Tribune* is apparently available digitally, online, and who knows, maybe a Gil Longworth and two guards were killed by Indians during this period and Norvill found the story. Still, that's a long way from saying they just dropped off a gold shipment to the skipper of a river steamboat. But my guess: the whole story with the exceptions of what we know from reliable and historical sources is pure fiction.

Marvin Brooks

Sources

Pepper, Choral: *Treasure Legends of the West*, Gibbs Smith (1994)

Norvill, Roy: *The Treasure Seeker's Treasury*, Hutchinson & Co. (1978)

Wilson, Ian: *Undiscovered: The Fascinating World of Undiscovered Places, Graves, Wrecks, and Treasure*, William Morris (1987)

Schurmacher, Emile: *Lost Treasures And How to Find Them!*, Coronet Communications (1968)

Frost, Lawrence: *Custer Legends*, Bowling Green Univ. Popular Press (1981)

Deem, James: *How to Hunt Buried Treasure*, Houghton Miffin (1992)

Hanson, Joseph: *The Conquest of the Missouri*, Murray Hill Books (1946)

Amaral, Anthony: *Comanche: The horse that survived the Custer massacre*, Westernlore Press (1961)

EMails

Davis, James A.: Head of References, State Archives, State Historical Society of North Dakota

Scott, Dr. Douglas: Battlefield Archaeologist & University Research Scientist

Researching the Missing Gold from the Steamboat Far West Story

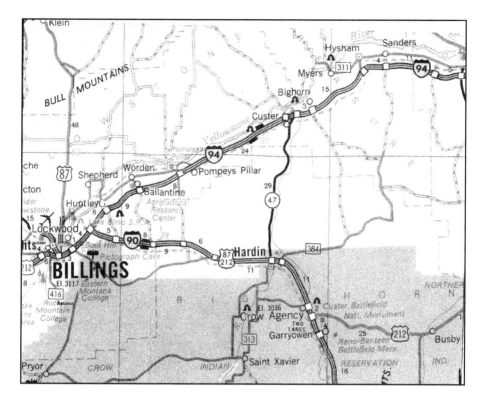

According to Pepper, the search area for that missing gold is in the hills near the river, up about fifteen miles from Hardin.

Researching Maximilian's Lost Treasure Story

In the previous four treasure stories we have touched base in New Mexico, Nevada, Colorado, and Montana. This next story is centered in Texas, way out in West Texas, at a place called Castle Gap. If true, this could be the biggest treasure of all, as far as land-based ones, for the value would be priceless. I admit that I had some misgivings about even including it in this book, for a couple of reasons. For one thing, if the story has any validity at all, the search for that treasure should be left to the archaeologists rather than the treasure hunter, but I know of no evidence that the story is taken seriously by those people. Then think about it, according to the story, the wagon train carrying Maximilian's treasure included forty-five barrels, filled with gold and silver and jewels, you know, that kind of thing. Why, it's just mind boggling. But let's suppose that the story is true, and some lucky treasure hunter located it, and maybe dug it up. Well, that old boy's troubles have just begun. To begin with, Texas has one of the toughest antiquities laws in the nation. If one removed any of that treasure, they would be subject to serious jail time, and if they reported the find to authorities, we can be sure some judge would give the anthropology department from some major Texas university full authority to dig and catalog all items recovered. Then the lawsuits would begin. In addition to federal, state, and county governments who would claim ownership of the spoils, there would be the landowner of the property who would likewise claim ownership. Then there is both Mexico and France who would claim it was their property to begin with. And let's not forget Maximilian's famous family, the Hapsburgs; they would surely claim ownership. And that poor old treasure hunter, well, if not in jail, he sure wouldn't get any treasure. And all that gold and jewels and stuff would probably end up in some museum. But just as I pointed out in my story on Forrest Fenn's treasure, most treasure hunters, including myself, usually take the attitude, what the heck, I'll find the treasure first, then I'll worry about the legalities.

But I'm telling you, it's getting tougher all the time being a treasure hunter. I remember many years ago being at a beach pavillion on the north end of Padre Island, close to Corpus Christi. It was wintertime and absolutely no one was there. So I had my metal detector out looking for lost items from the previous summer. A Federal Park Officer arrived. Let me tell you he wasn't nice at all—chewed me out good—told me

Researching Maximilian's Lost Treasure Story

he could confiscate that detector on the spot and told me to get off that federal property now. I was on the Padre Island National Seashore. Yes, "Treasure Hunters" must steer clear of federal parks and property. But let's get back to the story at hand, you remenber, Maximilian's lost treasure, said to be buried out there in West Texas. Where did that story come from?

Well, I have two versions before me at this time. One is from Choral Pepper's *Treasure Legends of the West*, the same book that started me on the research for "Custer's Gold." The other is a shorter version from probably the most famous of all treasure books, (well, except maybe *Treasure Island*), J. Frank Dobie's *Coronado's Children*. In summary, here is the story: Maximilian, an Austrian from the famous Hapsburg family, along with his beautiful young wife Carlota, were placed on the throne in Mexico by Napoleon III, as Emperor and Empress. It was a bad gig for these two. After a few years Napoleon, who had troubles elsewhere, withdrew the French forces from Mexico, leaving Maximilian very vulnerable indeed to a civil war led by Benito Juarez. The writing was definitely on the wall, and there is a lot of speculation as to why Maximilian did not abdicate his "throne." Carlota had the good sense to get the hell back to Europe, supposedly to try and persuade Napolean to protect the "Empire" by keeping French forces in Mexico. He was not persuaded and Carlota never went back to Mexico. So why didn't Maximilian escape also?

Now I want to make it clear that I am not an historian, not even an amateur one, and I have not studied the tragic story of Maximilian and Carlota very closely, but from reading about their last days in Mexico, it appears that even the historians disagree about Maximilian's motives in his last days in Mexico, before he faced a firing squad on the orders of Juarez. I've seen speculations that his mother encouraged him to stay and fight, rather than flee, and dishonor his family's name. Others believe that he was an honorable man and chose to lead his followers in a hopeless battle rather than abandon them. But I will leave all that to the historians. Nevertheless, the ultimate fate of Carlota held my interest. Shortly after returning to Europe, Carlota went, as they say, mad. She also became one of the richest people in Europe, through the efforts of her family. She lived to the year 1927 and died at the age of eighty-six, about sixty years after becoming Mexico's last Empress. Now back to the story of Maximilian's lost treasure.

It is thought that Maximilian brought a large private fortune with him to Mexico. What happened to it? Well, as this story goes, a band of six Missourians, ex-Confederate soldiers, on their way to Mexico to escape from life under the Carpetbaggers, met a caravan of wagons coming out of Mexico. The travelers with the wagons numbered fifteen, including the leader, an Austrian and his beautiful young daughter, other Austrians, and Mexicans. The leader indicated a concern about road conditions up to San Antonio. After the travelers revealed that the road was extremely dangerous due to hostile Indians and bandits, the leader, indicating that he had a valuable load of flour, offered the six Missourians a job as guards, for good pay, which they accepted.

Now on their way, the ex-soldiers became suspicious of the nature of the cargo, noticing how closely it was guarded. One of the six was chosen to sneak a look into one of the barrels. He was sucessful and reported it was loaded with all sorts of valuable coins, vessels of gold and silver, and gold bullion too. The six immediately began planning to kill the wagon crew and steal the cargo.

As the caravan approached the Pecos River, and the pass known as Castle Gap, the six Missourians constituted the watch while the others slept. All fifteen of them, including the young girl, were slaughtered. Their bodies were burned along with other properties. But, as Dobie tells us in his account of the story, "Papers taken from a chest revealed that the leader of the dead band was one of Maximilian's followers entrusted with carrying the royal fortune out of Mexico to Galveston." From there it was to be shipped to Austria where the Empress Carlota awaited, and where Maximilian, fleeing Mexico, was soon to join her, or so it was thought, according to the story.

Well, these six cutthroats didn't want to take all that cargo into a settlement, thereby drawing attention to themselves, so they decided to take what coins they needed to satisfy immediate desires, bury the treasure and come back for it at a later date. Now on the way to San Antonio, at Ft. Concho, one of the six became ill, and had to be left behind. Lucky for him. Very shortly, the other five were massacred by Indians. After recovering, the lone survivor headed out to San Antonio and eventually learned his partners were dead. Yippee! He was now the sole owner of Maximilian's fortune.

But things didn't work out too well for him. He decided to go to Missouri to get help from the James boys, Jesse and Frank, friends of his, to help him recover the treasure (this seems a little far-fetched). But he never made it. As Dobie tells us, he teamed up with some horse thieves and got arrested along with them. Now in the Denton, Texas, jail, he again became deathly sick. Told by a local doctor that he could not recover, he sent for a lawyer by the name of O'Connor, and turned over to him, and his doctor, a Dr. Black, a plat (which is another name for a map) to the fortune buried at Castle Gap, told them all the circumstances connected with it, and then "gave up the ghost."

When Black and O'Connor got out to Castle Gap, they learned that the "terrific" sandstorms common to the region had "shifted the landscape." All the two of them found were some wagon irons marked by fire. Therefore, according to Dobie, "Sphinx-like in its muteness amid the deep and solemn sands, Castle Gap still guards Maximilian's gold."

Well, it's just a great lost treasure story. The two versions by Dobie and Pepper are basically the same as far as pertinent information, but Pepper's version is dressed up a lot more with details about the life and times of Maximilian and Carlota, and was published in 1994, sixty-three years after Dobie's version, published in 1930. Pepper offers two other references in addition to Dobie's classic book. I have possession of one of those, *Maximilian and Carlota: A Tale of Romance and Tragedy*, by Gene Smith, published in 1973. I skimmed through most of it and read some of it having to do with the last days of these two tragic figures in Mexico—Carlota's escape to Europe, Maximilian's last battle, his capture by Juarez's forces, and his execution. I also read some about

Carlota's long life after fleeing Mexico. It is a compelling story, and I'm glad I learned something of it. But I was looking primarily for accounts of Maximilian's "fortune," and whether or not there was any mention of it, especially anything to back up Dobie's and Pepper's stories. I found nothing in that regard, but I did take note of passages having to do with Carlota's departure from Mexico. Unannounced to anyone, Carlota headed to Vera Cruz where a French ship awaited her. Maximilian accompanied her to the Shrine of the Virgin of Guadalupe (wherever that is) then headed back to Mexico City to face his fate versus Benito Juarez. Carlota headed on east, and as reported by Smith, "She carried with her for expenses thirty thousand dollars taken from the special funds set aside to fight flooding in the capital. It was the last substantial sum the Empire owned." So it doesn't appear the Emperor and Empress had such a tremendous fortune to be sent up to Galveston. Nor does it appear that Maximilian had any intentions of fleeing Mexico.

So at this point in my research I had Dobie's and Pepper's versions of the buried treasure at Castle Gap, one preceding the other by sixty-three years. I had a third reference from Pepper, *The Treasure Legend of Castle Gap: A collection of treasure legends*, by Jeff S. Henderson, published in 1965, and as I write this I have it ordered as an interlibrary loan. Unless I find something in that book to change my mind, I will conclude that this treasure story is indeed just a legend, and one that was originated by an old writer of legends, a legend himself, J. Frank Dobie. Now why would I believe that? Well, Dobie doesn't offer a bibliography in his book to provide us with a source of his story, but he does offer notes. In regards to the story at hand he says, "This legend has been supplied by J.A. Rickard of O'Donnell, Texas, who heard it from a frontiersman named T.J. Kellis." So we are getting this treasure tale third or fourth hand, and without an answer to a very pertinent question: Where did that frontiersman get his information? Maybe I will learn some more details when I receive Henderson's book. It's on its way.

And I now have it in hand. Well, I was a little disappointed. I guess I expected a well-researched tome telling me in great detail all about that treasure of Maximilian, buried all these many years out there at Castle Gap. But the book is only forty pages of which nine are illustrations. In that limited space Henderson writes of four treasure stories connected to Castle Gap, but clearly features the Maximilian treasure. I'll get to his version of the story but first I want to point out that, in his introduction, he tells us that the place known as Castle Gap lies just east of present-day Highway 385 and provides a break between two mountains, Castle and King Mountains. (I've provided a map). The Gap is a little more than a mile in length and about three-quarters of a mile in width at its widest. According to Henderson, it has seen raiding Indians, patrolling soldiers, stagecoach passings, emigrant wagon trains, cattle drives, and horse wranglers. Then he tells us that in recent times, "The silent old canyon has chuckled in the soft breezes that whistle along its walls at the many treasure hunters that have sought the hidden fortunes buried somewhere within its confines." Okay, but what does he have to tell us about that Maximilian treasure?

Well, as it turns out, not much. In fact, Henderson quotes Dobie's story almost verbatim, but gives credit to Dobie. Then after finishing with what he calls "Dobie's yarn," he tells of several versions of it; for example, that the treasure was from the Church of Mexico; or that the Maximilian treasure actually exists but is lost in other locations, possibly at sea. Most of these alternative stories appear to be tales of folklore, passed on from one generation to the next. I did not find evidence of any serious studies to verify that this treasure ever existed, nor is there any record of a first person account by Dr. Brown, that dying mass murderer's final doctor, or his attorney, Mr. O'Connor, that shares with us what must have been a grisly story, that being the cold-blooded killing of fifteen people, including the leader's beautiful daughter. In absence of any of that, folks, I think I will give J. Frank Dobie, a masterful folklorist, credit for the tale, as told to us in his 1930 classic, and believe him when he tells us he got the story as handed down from a frontiersman, you know, the one named T.J. Kellis.

But here are some unanswered questions: Did Maximilian and Carlota bring valuables to the New World? That seems likely, for they were known to live in high style. If so, then, is there any evidence of what happened to those valuables? I don't profess to know the complete answers to those questions, but in my research I found a clue. It was in another book on Maximilian that I looked into to see if I could find anything to substantiate the tales put forth by Pepper, and apparently originating with Dobie. The book is *Maximilian and Carlota: Europe's Last Empire in Mexico* by M.M. McAllen, published in 2014. Benito Juarez, on his death, was succeeded by Porfirio Diaz, who ultimately served as president of Mexico for over thirty years. Apparently Porfirio was influenced by the high living standards of Maximilian and Carlota. As McAllen writes:

> During his grandiloquent tenure, he maintained rooms with possessions from Maximilian's reign, including the silver, furnishings, pianos, artwork, even preserving the coaches, the lavish ceremonial vehicle and the utilitarian one. He held "audiences" in the manner of the Second Empire, in the same rooms of the National Palace, and people often commented on his elaborate state dinners for which he used Maximilian and Carlota's crystal, silver, and china.

McAllen's 509-page tome appears to be well researched. She has ninety-eight pages of notes and bibliography. She makes no mention of any of Maximilian's possessions being carted off toward Galveston, Texas, to be shipped to the waiting Carlota. No, after due consideration, I must conclude this whole tale orginated in folklorist J. Frank Dobie's 1930 book, *Coronado's Children*, and it was all started, apparently, by that frontiersman T.J. Kellis. Now, did Kellis make the story up, or was he passing on one told to him? Who knows? But there are indications that "treasure hunters" even today, 150 years later, are still traveling out to a place called Castle Gap, Texas, thinking they just might find all that gold and silver. And that's not unusual—lost treasure stories seem to affect some of us that way.

Sources

Pepper, Choral: *Treasure Legends of the West*, Gibbs Smith (1994)

Dobie, J. Frank: *Coronado's Children*, Hammond, London (1930)

Smith, Gene: *Maximilian and Carlota: A Tale of Romance and Tragedy,* William Morrow (1973)

Henderson, Jeff S,: *The Treasure Legends of Castle Gap: A collection of treasure legends*, Mayo Cleveland (1965)

McAllen, M.M.: *Maximilian and Carlota: Europe's Last Empire in Mexico*, Trinity University Press (2014)

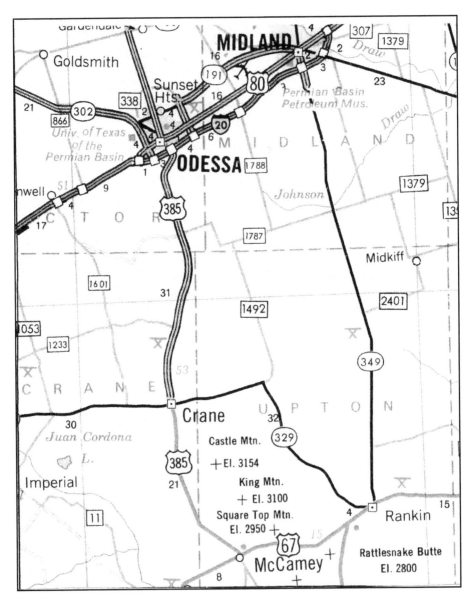

Castle Gap—only a mile in length—is a pass between Castle and King Mountains.

Conclusion

It has occurred to me that if this book sells lots of copies to serious Fenn treasure seekers, there might be scores of people down on the La Madera trails, each holding a copy of my book to "guide" them. Maybe it was Fenn's concern about human traffic that caused him to warn a successful hunter to be quick, and not to "tarry," gazing about. How would one secrete that forty-two pound chest? You know, I worry about little details like that. I would say if one has a good sized pack, that would do the job. But remember, I said I had an active imagination. I can envision people hurrying down those canyons and trails in a frenzied rush to locate the blaze and beat their fellow seekers to that virtual pot of gold. I've thought that some enterprising folks might set up a lemonade stand out there on the trails, to make a little cash, and I suppose someone has already thought about printing up a tout sheet, to sell at the news stands in Santa Fe, or on the streets, providing all of those 30,000 treasure hunters the latest tips on locating old Fenn's treasure chest. How well I remember that great movie comedy, "It's a Mad, Mad, Mad, Mad World." It showed how people can react to "gold fever," or the prospects of finding treasure. Well, I just hope nothing like that happens in northern New Mexico as a result of my book.

But to get serious one more time, I firmly believe that no one is going to get very far with Fenn's clues, without taking a few of those "leaps in faith" as I did. Sure, it might be that the treasure is not in northern New Mexico, and even if it is, it might not be close to the Chama River. And my "logical, deductive" reasoning in focusing on those two particular clues, concluding that "the wood," refers to La Madera, or that "the blaze," refers to markings or signs in "the wood," leading one finally to the treasure, might well be totally "off track." But I believe it offers a better chance on success than trying to make sense out of each of Fenn's nine clues, provided even that one can find those. My idea on the meaning of the nine clues is as good as any I've seen. And I wasn't entirely joking when I said I wouldn't mind going down to northern New Mexico and checking out those "solutions" I've come up with. You readers better hurry. I might beat you to that gold-laden chest (just kidding). But I know that I have gone as far as I'm going to regarding Fenn's clues, and frankly, I'm kind of relieved. Reading and rereading Fenn's poem and trying to make sense out of it can become nerve-racking.

As to the second treasure story in this book, on the Steamboat *Gila* robbery–Crescent Springs treasure tale, I think I've made it clear that there were a number of things in that story that intrigued me. I found it amazing that at least nine, and maybe more, authors wrote articles on that story without once questioning its authenticity. I suppose most of them just wanted to get a little treasure story published.

I had written up an earlier version of my research efforts on that story, and gave a copy of it to a friend. He said he found it interesting, but started to feel sorry for me toward the end, obviously referring to my difficulty in solving all the mysteries in that tall tale. I don't think he realized that all that research was done as a pastime over about a two-year period, and that I found my research on the *Gila*, its captain, the Eldorado Canyon, and the story of the robbery itself of sufficient interest to keep going. Then along the way I encountered such interesting characters, real people, that some way found themselves in the tale: Wharton Barker, "Humbug Bill" Frazee, his newspaper, the *Green-eyed Monster*, and the boys at McGintey's, Don Ashbaugh, Foster McClure, and old Ike Allcock. I don't regret at all spending my time trying to unravel all the mysteries that author Fred Kuller threw into his farce. I just regret that I wasn't able to learn anything about him. I'd sure like to know how he learned about all the real people, and real places, he put in his tale. I hope you readers, who probably bought this book because of an interest in Fenn's treasure, will nevertheless enjoy my research into this second treasure tale. If nothing else it shows how a legend is born.

The third story, on the missing Denver Mint dimes, was one that got my interest when I saw it first mentioned in one of Thomas Penfield's treasure "guides," perhaps because it concerned a Colorado treasure. I learned that there are many people, even today, discussing that treasure tale, and even talking about coming to the Black Canyon, in search of those kegs of dimes, just waiting there to be picked up by a treasure hunter brave enough to go down there in that fearsome-looking canyon. Then I was very lucky in finding "Apache Jim's" book in my local library. I think it's clear that, with the exception of Jim's story itself, there is absolutely no evidence of any dimes down in the dangerous-looking Black Canyon. However, I enjoyed Jim's story and researching it, and I hope you readers find it of interest.

If the purpose of my research into the four treasure tales was to either debunk or authenticate the story, then my efforts on the missing gold shipment from the *Far West* were probably the most frustrating. It just seems doubtful that that famous treasure story, involving real people, real events, and a real steamboat and her captain, could have originated in a story written by a British author, Roy Norvill, and published in 1978. Yet I could find no evidence of any earlier version of his tale. Certainly, there was nothing in Emile Schurmacher's book to authenticate his tale, as Norvill would have us believe, and I found no evidence whatsoever that Schurmacher version, told in the third person without any references, was anything more than that of Norvill—an example of that genre known as "historical fiction," which I talked about in length in my story.

But there was one book having to do with Custer's battle and the role of Capt. Marsh, and the *Far West* that was definitely serious. That was Joseph Hanson's *The Conquest of the Missouri*, published in 1946. My library's Special Collections has a copy of that book, and I skimmed through it, early on, to see if there was any information in the book to verify Pepper's story, and found nothing. But after finishing up my own story, and not being totally satisfied with my conclusions, I decided to revisit that book.

I learned, on closer examination, that the book was first published in 1909, and was reprinted in 1946. The book is basically an account of the adventures and achievements of Grant Marsh, a pioneer of steamboat travel on the Missouri and Yellowstone Rivers, of which there are many. But I focused on that period having to do our treasure story, to be sure that I didn't overlook some details that might lend credence to Pepper's, Norvill's and even Schurmacher's tales. I found no reference to anything having to do with a gold shipment, nothing at all to authenticate these tales. Further I found no evidence that the *Far*

West engaged in any battles with Indians on the way to Ft. Lincoln with the wounded soldiers. I did find reference to some of the names of *Far West* crew members including Mate Thompson, Eng. Foulk, and Pilot Campbell. And this makes me wonder if some of these writers didn't have access to Hanson's book. But if they did, they gave him no credit.

So I found Hanson's book something of a classic, and well worth looking into for anyone interested in what happened up there during the period in question, but no, I found nothing to change my conclusions regarding this particular treasure tale.

Finally, there was the Maximilian fortune–Castle Gap treasure tale. You know, I've been thinking a little more about that "yarn." That must have been quite a hole those cutthroats dug out there, to be able to bury forty-five barrels of treasure. Why, they didn't even have a bulldozer. But I guess they had plenty of shovels to do the job. I wonder if they didn't worry that someone would come through the pass and catch them red-handed in their evil deeds. But, as far as my research, I'm satisfied that that particular story came from the pen of J. Frank Dobie, and is simply one of hundreds that he wrote throughout a long life. It was so insignificant that he only devoted a very few pages to it in his classic lost treasure book, *Coronado's Children*.

I know that you readers have taken notice that I have debunked all the treasure stories I researched for this book, with the exception of Fenn's hidden treasure. I think I pointed out before that I didn't set out to do this—I really wanted to find credibility in the stories so that I might go out and try to find one of them. But there is no way to get around it, those four treasure stories I'm referring to just don't pan out, as an old gold prospector might say. But then the question to me might be—Well, do you know of any lost treasure stories that you believe can be authenticated? I have an example of one, but before I get into it, I want to point out that there are literally hundreds of lost treasure and lost mine stories in the state of Colorado alone, including two books I am familiar with that are cram-packed with such tales. Most of them are so lacking in detail that I am not motivated to spend any time researching them, but here is

Conclusion

something more. I don't recall ever seeing any evidence that anyone has actually located one of these treasures. I know, I know, maybe they kept their find secret, but still, as far as land-based lost treasure stories, one must admit, the evidence is pretty scant that these "treasures" are anything more than tall tales, as I think proved to be the case with the four stories I researched for this book. But there is one more story that I thought about researching for this book. I feel certain the story is authentic. But my problem was the time and travel it would take to pursue this treasure, or to be more precise, these lost treasures—there may be numerous ones. Call this story "The Lost Treasures of Bernarr Macfadden."

Macfadden was born in 1868 and died in 1955, at age 87. I will refer the reader to the internet for details of his life, but he was a very successful publisher, and an eccentric one. For one thing, he didn't trust banks, and that particular quirk is the basis for my interest in him. From the reports I have seen he buried his money, and possibly other valuables, in metal ammunition boxes in a number of locations on his properties. As in most well-known treasure stories there is a wealth of information on Macfadden's treasures on the internet, and I recommend those to any reader interested in this story, but I believe the best information can be found in a book published in 1962 by his widow, Macfadden's fourth wife, Johnnie Lee Macfadden, entitled, *Barefoot in Eden: The Macfadden plan for health, charm, and long-lasting youth.*

Well, who could resist a title such as that? I was able to borrow a copy of this book through the interlibrary loan system, and I found it to be a pretty charming little story about Johnnie Lee's life with Bernarr. Of course, she was decades younger than him, and she relates what it was like living with this extremely eccentric man in the twilight of his life. So I recommend the book to anyone interested in Macfadden's "plan," and his life. But I became aware of the book because someone, somewhere pointed out that Johnnie Lee relates, in her book, about Macfadden's habit of burying his money. Indeed she does; she wrote a full chapter entitled, "Buried Treasure."

Johnnie Lee talks a lot about the careless way Bernarr handled his money, relating one instance where he accused her of taking his roll of bills. It was later found in his shoe and Johnnie Lee estimates the roll contained about $5,000. Some time after her marriage, Johnnie Lee learned that Bernarr buried his money, and here is a quote from her book, "Over the years, in moments where he was feeling particularly affectionate, Macfadden confided in me that he had buried about four million dollars in various places all over the country, so that no matter where we might be, he would be near a source of ready cash." For awhile Johnnie Lee thought this was all a fantasy, but she awoke one morning finding him missing, went looking for him, and found him digging a hole to hid his ammunition box. This infuriated Bernarr, and he made it clear she was never to follow him again. He told her, in anger, not to be sneaking around after him, then told her, "Some day before I die, I'll give you a map, but I'm not going to die for another fifty years, so don't get anxious." Bernarr believed he would live to be about 150.

Johnnie Lee ends this chapter by telling us that in 1961 a bulldozer near Jericho, Long Island, dug up a metal box measuring 18 by 10 inches. Inside were bills wrapped in neat packages in perfect condition totaling $200,000. She says the money was "immediately taken under official jurisdiction," and finishes with this line, "His estate has been tied up in endless legal battles, and so far I have never received a cent from the estate of my multi-millionaire husband." So Johnnie Lee's book was published 1962, seven years after Bernarr died. Did she ever receive anything from his estate? Did she ever locate that map Bernarr talked about? Have there been any reports that other Macfadden treasures have been found? I don't know the answer to any of those questions.

A *New York Times* obituary reported that Johnnie Lee Macfadden, author and physical fitness advocate, died on April 6, 1992 in her home in Manhattan. She was eighty-eight years old, and survived by a son, a daughter, and a number of grandchildren and great grandchildren. Her book is available on Amazon, as are numerous books by Macfadden.

Johnnie Lee mentions several places Macfadden lived during his life; places the two of them lived while married, and even places where he might have hidden his money. She talks about a hill near his hotel in

Conclusion

Danville, New York; along the Palisades, on the cliffs above the Hudson River, near Nyack, New York; a residence in Far Rockaway, New York, eighteen miles from where the bulldozer dug up his treasure near Jericho, Long Island; and she says he had a palatial estate in Englewood, New Jersey. He owned the Deauville Hotel in Miami Beach, and the Arrowhead Spring in California, and remember, Macfadden had had a long life before marrying Johnnie Lee, fathering nine children with several wives. Who knows how many treasures he hid over his long life, or in how many locations he hid them.

Finally, there is this, a quote from Johnnie Lee: "Actually I never really have tried to find any of the buried treasure, considering it useless just to start out and dig with no real clue as to the various locations of the hidden money." Obviously this woman doesn't have the imagination or dreams of the normal Forrest Fenn treasure hunter, or the tenacity of many metal detectorists. Otherwise, she would have been out there on those properties she talks about, from the death of Bernarr, until her dying day, scouring the woods with her metal detector, or going through every document or piece of paper left by Bernarr, looking for that treasure map. But that's just me, an old "treasure hunter" talking. Still, I wonder, did Bernarr actually have a treasure map, and if so, where is it? It might be worth millions.

So, yes, I believe there are lost treasures out there, and Bernarr's is one of them. But it will require a whole lot of research to have any possible chance of zeroing in on one of those locations Johnnie Lee talks about. Still, the evidence is clear, Bernarr's hidden treasures are most likely still out there, just waiting for that persistent treasure hunter to find them. Good luck hunting.

CPSIA information can be obtained
at www.ICGtesting.com
Printed in the USA
BVHW01s1240010318
509345BV00031B/419/P